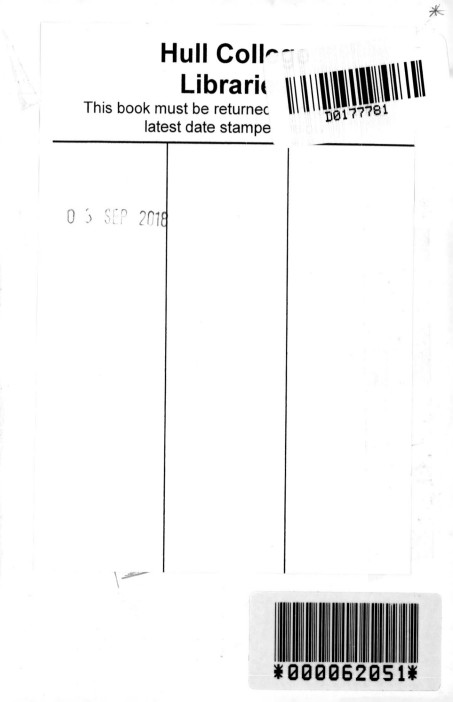

Frontispiece: Limestone showing the tube-like traces of burrowing animals. Port Issol, France.

THE MACMILLAN FIELD GUIDE TO
GEOLOGICAL STRUCTURES

John L. Roberts

MACMILLAN

First published 1989 by Macmillan Press.

This paperback edition published 1996 by Macmillan Reference Books
a division of Macmillan Publishers Limited
25 Eccleston Place London SW1W 9NF
and Basingstoke

Associated companies throughout the world

ISBN 0 333 66295 4

1 3 5 7 9 10 8 6 4 2

A CIP catalogue record for this book is available
from the British Library

Typeset by Glyn Davies, Cambridge

Printed in Spain

CONTENTS

Preface

Geology is still essentially a visual science, much concerned with the natural appearance of rocks in the field. Indeed, the very foundations of the science were first established, some two centuries ago, simply as the result of such observations. The first great steps in understanding the history of the earth were taken at that time, once it was realized that rocks should not just be studied in isolation, but that it was even more important to understand how they "fitted together" on a much larger scale in the field. In fact, once it was understood how different rocks occur in contact with one another, which is the essence of geological structure, a much clearer idea of their origins emerged.

The present book provides an introduction to the wide variety of geological structures, concentrating in particular on the structural features which are commonly enough displayed by rocks in the field. However, rather than just describing the minutiae of geological structures as seen in individual exposures I have deliberately widened the discussion in the text to provide an appreciation of geology as a whole, as far as this is based on an understanding of the structural relationships which can be seen in the field. I hope that this approach will interest the amateur naturalist without any great knowledge of the subject, as well as the professional student of geology, studying the subject in college or university.

Geology, like most other descriptive sciences, has acquired a formidable terminology over the years, often abandoning the vernacular in favour of complex Latin and Greek roots for its key-words. This provides a major obstacle to its study, at least to the amateur. Fortunately, the study of geological structures in the field is affected less by this difficulty than other branches of the subject. Even so, I have tried to be scrupulous in defining all the technical terms likely to be unfamiliar to the general reader when first introduced into the text, giving their derivation from Latin or Greek wherever this casts some light upon their meaning.

The reader will notice that I have rarely used a geological hammer as the traditional method of illustrating the scale of geological photographs. Although I hope that anyone interested will be encouraged to visit the localities listed to see the evidence at first hand, it is emphasized that **exposures should never be damaged by hammering**. Reasons of conservation suggest that the time has now come to abandon the use of geological hammers particularly by field-parties, for the benefit of future visitors. Too often, indiscriminate hammering has simply destroyed the delicate features shown by rock exposures, weathered out over thousands of years, so that they are now lost to us forever.

Acknowledgements

I could not have written this book without spending many days in the field in the company of all my geological friends and colleagues, to whom I am therefore much indebted.

My thanks are due to Gill Harwood and Colin Scrutton for providing the photographs as listed in the captions. Bert Randall presented me with a gift of Tomkieff's *Dictionary of Petrology*, for which I am most grateful. I would also like to thank Professor J.E. Hemingway, who kindly acted as a friend and mentor while I was at Newcastle. This is also the place to acknowledge my great debt of gratitude to Jack Treagus for his friendship over all those years since we started working together in the Scottish Highlands. Finally, I would like to thank my wife, Jessie Roberts, for all the help and support that she extended to me while working on this book. Lacking a scientific background, she has taken the trouble to read through the text to identify any particular points of difficulty for the lay reader. However, I still remain responsible for all errors of fact and interpretation.

Author's Note

Duplicate sets of the 35mm colour transparencies used to illustrate this field-guide are available for purchase, together with 20 × 30 inch photographic prints of individual slides. Please write for further details to Geological Field Photographs, Clar Innis, Strathtongue, Tongue by Lairg, Sutherland IV27 4XR, Great Britain.

Photographic Techniques

The transparencies used to illustrate this field guide were mostly taken under natural light using an Olympus OM-1N camera, fitted with Tamron SP 35–80mm and 70–210mm zoom lenses. Fujichrome 50ASA and 100ASA film were used for preference, with the 100ASA film being exposed at 200ASA for push-processing at the laboratory. Exposures were generally taken at minimum aperture to give maximum depth of field. A polarizing filter was used occasionally to reduce reflections from water.

Photographs lacking a scale are mostly of inaccessible exposures, taken using a telephoto lens; any statement concerning the scale of the photograph is an estimate. The Tamron lens-cap used as a scale has a diameter of 6.5cm, just over 2½ inches. Where coins are used, the one-penny piece has a diameter of 2cm, just over ¾ inch, while the two-penny piece has a diameter of just over 2.5cm, or 1 inch.

PART I
SEDIMENTARY ROCKS

Eras	Periods	Age Yrs (millions)	Orogenies
Cenozoic	Quaternary	2	Alpine
	Tertiary	65	Laramide
Mesozoic	Cretaceous	144	Nevadian
	Jurassic	213	
	Triassic	248	
Palaeozoic	Permian	286	Alleghenyan Hercynian
	Carboniferous	360	
	Devonian	408	Acadian
	Silurian	438	Caledonian
	Ordovician	505	Taconic
	Cambrian	570	Grampian
Precambrian Times			Many Orogenies
Origin of the Earth		4600	

Table 1 *Stratigraphic Time-scale.*

4

NATURE OF SEDIMENTARY ROCKS

Mode of Origin

Sedimentary rocks are formed from loose material, mostly derived from the weathering and erosion of pre-existing rocks, which is deposited (Latin: *sedere*, to settle) on the Earth's surface. This material is usually carried away from its source through the action of running water, by rivers and streams on the land, until it reaches the sea, where it is transported across the sea-floor by waves, tides and other currents. Locally, the wind is an important agent of sedimentary transport, particularly in deserts, while glaciers also carry along sedimentary material as they flow downhill. All these agents of sedimentary transport become the medium for sedimentary deposition once they are unable to carry their load of sedimentary material any further. However, water is the most important agent of sedimentary transport, and therefore of sedimentary deposition. Many of the characteristic features of sedimentary rocks are therefore a result of their deposition under water.

Weathering and erosion produce sedimentary *detritus* (Latin: *detritus*, worn away) composed of rock fragments, mineral grains and clay particles (terms given in order of decreasing grain-size). As this is transported, particularly by running water, it tends to be sorted into fractions of differing grain-size. Each fraction is deposited once there is no longer sufficient energy for its further transport. Deposition typically occurs in a number of different environments, allowing sedimentary rocks to be distinguished according to their *facies* (Latin: *facies*, form or appearance) which is a subject well beyond the scope of this book.

Rock fragments and any mineral grains greater than 2mm in diameter are usually deposited first to form *conglomerates* and *breccias*. Conglomerates consist of well-rounded pebbles (2–8mm), cobbles (8–256mm) and boulders (over 256mm), whereas breccias (Italian: rubble from broken walls) are formed by more angular fragments. Any finer-grained material composed of particles less than 2mm in size, but more than 1/16mm, is known as sand, whatever its mineral composition. However, much sand is formed by quartz as it is particularly resistant to the effects of chemical weathering, which breaks down the minerals in the original rock under the influence of the atmosphere. Quartz grains, and any other mineral grains or rock fragments of the same size, accumulate to form *sandstone*. A number of different varieties are recognized. *Arkose* (? Greek: *archaios*, ancient) is a coarse-grained sandstone containing much feldspar in addition to quartz, while *greywacke* (German: *grauwacke*, grey rock) is a poorly-sorted and often dark-coloured sandstone, rich in feldspar grains and sand-sized fragments of igneous and metamorphic rocks, set in a much finer-grained matrix of clay minerals.

Next, there are silt-sized particles, varying between 1/16mm and 1/256mm in diameter, which form *siltstone*. Finally, clay minerals occur as extremely small particles, less than 1/256mm in diameter. Unlike the coarser-grained detritus already described, which consists of mineral grains and fragments from the original rock, the clay minerals are formed by chemical weathering. They are eventually laid down to form *shale* and *mudstone*. All these sediments represent the clastic (Greek: *klastos*, broken into pieces) rocks, otherwise known as the detrital sediments.

Chemical weathering also produces material in solution, which is eventually carried away into land-locked lakes and, more commonly, the sea. There, it accumulates by evaporation, so forming sea-water. If evaporation is sufficiently intense, this material comes out of solution as chemical precipitates, so forming salt deposits, which are known in the geological record as *evaporites*. However, living organisms may also abstract this dissolved material to form their shells and skeletons, which can then accumulate as sedimentary material after their death. Rocks formed in this way include *limestones* and *dolomites*, composed respectively of calcite and dolomite, together with the *cherts*, formed by extremely fine-grained quartz. Chemical processes as well as organic activity are commonly involved in the formation of such rocks. Other sedimentary rocks, belonging to this very broad category of the *chemical-organic sediments*, include ironstones, phosphates, alumina-rich bauxites and the coals.

Compaction and Consolidation

Geology does not make any distinction between the loose and incoherent condition in which much sedimentary material is deposited, and the hard and solid nature of the sedimentary rocks eventually formed by this material as part of the geological record. Indeed, any deposit of sedimentary material is regarded as a rock, whatever its physical condition. Even so, physical and chemical processes affect sediment soon after its deposition, tending to convert this loose material into what is known popularly as a rock. These processes of *consolidation* result in loose sediment becoming lithified (Greek: *lithos*, a stone). This occurs partly in response to *compaction*, whereby the sedimentary particles become more tightly packed together under the weight of the overlying rocks. The volume is reduced largely by compaction in a vertical direction, together with the expulsion of any interstitial water occupying the pore spaces between the sedimentary particles. At the same time, the sedimentary grains may become further consolidated by the deposition of mineral matter in the pore spaces as a result of *cementation*. Solution and recrystalization can also cause them to stick together even more closely. All these post-depositional processes acting together to form solid rock from a loose sediment are known generally as *diagenesis*.

Bedding and the Principle of Superposition

Sedimentary rocks typically occur as discrete layers called *beds*, which were deposited one on top of another to form *bedding* or *stratification*. (The latter term has the Latin word "stratum" as its root, meaning something that is laid down or spread out.) Although "stratum" is now considered obsolete as a synonym for a sedimentary bed, the plural "strata" is still used when referring to a series of sedimentary beds which are interbedded with one another to form a *stratigraphic sequence*.

The individual beds in a stratigraphic sequence are separated from one another by *bedding-planes*. These are the surfaces on which each successive bed was deposited to form the sedimentary sequence. The bedding-planes therefore mark successive surfaces of deposition, each forming the Earth's surface just before the next bed of sedimentary rock was laid down. This fact enables the age-relationships of sedimentary rocks to be determined according to the *Principle of Superposition*. This principle simply states that each bed in a sedimentary sequence is older than the overlying rocks, but younger than the underlying beds, since it would be impossible for such a bed to be laid down underneath another bed which had already been deposited.

It is commonly the case that each bed of sedimentary rock within a stratigraphic sequence does not vary much in thickness as it is traced over a wide area. Thus the sedimentary layer forms a distinct unit sandwiched between two other beds, above and below. The beds are then said to be *conformable* with one another, arranged in such a way that they succeed one another in a regular fashion, without any sort of structural break between the layers, at least on a large scale (see **1**). However, irregularities may be developed on a small scale, so that individual beds may vary somewhat in thickness, wedging out or merging with one another, while the intervening bedding-planes show random or even systematic changes in attitude as a result.

Sedimentary Beds and Bedding-Planes

Sedimentary beds can be distinguished from one another by changes in their *lithology*. This is a general term referring to the overall character of the rock, as seen particularly in the field. The lithological differences reflected in the bedding of sedimentary rocks are marked by variations in grain-size, texture, colour, hardness and composition, which allow the individual beds of sedimentary rock to be recognized as separate entities. Such differences in lithology are commonly accentuated by the effects of differential weathering and erosion, so that bedding is often seen to best advantage on weathered surfaces. The varying resistance of sedimentary rocks to such processes is also seen on a larger scale in the formation of landscape *features* such as escarpments and topographic "benches" marked by changes or breaks in slope, which are formed by the more resistant rocks.

The nature of the bedding seen on a large scale in the Jurassic and Cretaceous rocks of the Lebanon is shown in **1**. Although close to the horizontal, the bedding has been tilted by subsequent earth-movements so that it now dips at a few degrees to the left. The prominent features to be seen on this hillside have been produced by the differential weathering and erosion of the underlying rocks,

as the river has cut down into its valley. Note how individual horizons of sedimentary rock can be traced for considerable distances across the hillside without much change in their character.

The bedding-planes separating the individual layers of sedimentary rock are sometimes just planes of physical discontinuity in the rock, along which it will split apart. Such ease of splitting is termed *fissility*, especially where the planes of weakness are set very close together. It is characteristic of shales, allowing these rocks to be distinguished from mudstones which lack such a fissile structure. More frequently, however, bedding-planes are marked by actual differences in lithology between the layers, as shown in **2**.

Much can be seen by first viewing bedding at a distance. First impressions concerning the orientation of the bedding can then be confirmed on reaching the exposure, seeking out any differences in lithology if the bedding is obscure. If this is the case, great care must be taken to distinguish bedding from other structures such as joints (see **113**) and slaty cleavage (see **192**). Bedding can also be confused with the iron-staining which results from solutions percolating through the rock. This commonly produces a banding, as shown in **3**, which is not related to the bedding.

Figure 1 *Bedding in sedimentary rocks exposed on a large scale in the Lebanon. Three massive limestones form prominent cliff-like features, separated from one another by ground underlain by less resistant rocks. The overlying rocks are equally well-bedded, but on a much finer scale. Quartaba, Lebanon. (Height of section c 1000m)*

Figure 2 *Sedimentary beds and bedding-planes. The light-coloured beds are greywacke, while the darker horizons consist of shale. The shale forms thin partings of more fissile rock between the much thicker beds of more massive greywacke. Aberarth (SN 490649), Dyfed, Wales. (Height of section c 5m)*

Figure 3 *Liesegang "rings" along a joint-plane in sandstone, formed by the rhythmic precipitation of iron oxides from solutions percolating through relatively porous rocks. Such features can be mistaken for bedding, particularly where the iron staining mimics the appearance of cross-bedding (see* **10***). Seaton Sluice (NZ 344757), Tyne and Wear, England.*

STRATIGRAPHIC PRINCIPLES

The Principle of No Initial Dip

It is usually assumed that sedimentary rocks are mostly deposited as beds on a surface which was extremely close to the horizontal. This means that the Earth's surface had such a horizontal attitude wherever it was formed by the accumulation of sedimentary rocks in the geological past. Only if the sedimentary material was so coarse that it could not be carried any distance from its source in a particular environment under consideration, would the depositional surface have departed to any extent from the horizontal. The sedimentary rocks deposited on such a slope would then display what is termed an *initial dip*, lying at an appreciable angle from the horizontal.

The circumstances allowing sedimentary rocks to be laid down with an initial dip are somewhat restricted in their occurrence. However, initial dips up to 30° from the horizontal, and sometimes even more, are shown by scree deposits below cliffs, and by analogous deposits which form below coral reefs and around volcanic cones. Other examples result from the deposition of sedimentary rocks around buried hills, the building out of deltas into deeper water, and the formation of alluvial fans as rivers flow out from mountains on to low-lying ground. Initial dips occur in such cases simply because the sedimentary particles are much too large to be moved any farther, given the energy available for their transport.

Apart from such exceptions, sedimentary rocks are generally deposited with virtually no initial dip. This may be observed at the present day, since it is commonly found that sedimentary rocks are mostly deposited in environments such as the flood-plains and deltas of large rivers, the low-lying coastal plains and shore-lines surrounding the continents, and the shallow shelf-seas around the continents, together with the abyssal plains forming the floors of the oceans. The depositional surface is very close to the horizontal in all these cases. This is so because there is always sufficient energy available in these depositional environments to carry the sedimentary material far and wide, provided that it is sufficiently fine-grained, so filling up any irregularities in the depositional surface. It is this factor which results in the striking continuity shown by individual beds of sedimentary rock within a stratigraphic sequence, which may therefore be taken as another argument in support of the Principle of No Initial Dip.

The Principle of Strata Identified by Fossils

The study of *stratigraphy*, which is concerned with the nature of geological history, as determined from an examination of sedimentary sequences for the most part, rests on the Principle of Superposition. Indeed, this Principle is fundamental to the study of stratigraphy, since it allows sedimentary rocks to be placed in their *stratigraphic order*, with the oldest beds at the base and the youngest beds at the top of any sedimentary sequence. However, this Principle only dates sedimentary rocks with regard to one another, identifying a particular bed as older or younger than another bed in the same sequence. Even so, it provides the foundation for another great principle, which is essential to the study of stratigraphy. This is the *Principle of Strata Identified by Fossils*.

Sedimentary rocks often contain *fossils*, which are the organic remains of plants and animals (or their traces), incorporated into the sediment at the time of its deposition. It was William Smith (1769–1839) who first appreciated that the various formations of sedimentary rock, that he mapped out over much of England and Wales, could each be identified by the characteristic assemblage of fossils which they contain. It is now known that this is simply the result of *organic evolution* whereby particular species change into new and different forms, coupled with *organic extinction*, as a result of which other species die out completely, with the passing of geological time.

This means that each species is only present as a living organism over a certain span of geological time, before which it had not evolved from a preexisting form, and after which it had either evolved into a new form, or become extinct. Accordingly, each fossil species can be taken as diagnostic of a particular interval of geological time. This means in principle that sedimentary rocks can be dated stratigraphically by the fossils found preserved within them. The species which are particularly diagnostic of stratigraphic age are known as *zone-fossils*. The fossil record can therefore be used to construct a geological time-scale known as the *stratigraphic column*. This divides geological time into a number of distinct units, as shown in Table I. The ages assigned to the beginning of each stratigraphic period are given in millions of years, as determined from the results of *radiometric dating*. This method relies on the presence of radioactive isotopes, which are found to occur naturally in rocks and minerals, albeit in small quantities. These "parent" isotopes undergo radioactive decay with a particular "half-life", breaking down to form "daughter" isotopes, usually of a different element. Since the amount of any "parent" isotope gradually decays over geological time, while the abundance of the corresponding "daughter" isotope increases, it is theoretically possible to determine the radiometric age, if the "half-life" can also be measured.

Evidence for Earth-Movements

Although sedimentary rocks are nearly always deposited an horizontal layers, it is often found that the bedding no longer preserves such an attitude. It can be inferred that the rocks have been affected by *earth-movements* following their deposition.

4 shows a bedding-plane, exposed over a wide area, which has been affected in this way. What is then termed its dip and strike describes how it has been tilted away from the horizontal. Its *strike* is defined by the direction of a horizontal line, drawn at any point on this surface. Such a direction would be parallel to the high-water mark in the present case. It is measured by taking a bearing from true north with a compass. The *dip* of such a bedding-plane corresponds to its angle of greatest slope, down and away from the horizontal. It is measured at right angles to the strike, using a clinometer. The direction of dip should be specified as well as its angle, since a bedding-plane might well be inclined towards the left, when viewed along the strike in a particular direction, rather than the right.

Folding rather than the simple tilting of sedimentary strata is another result of earth movements, seen wherever the bedding shows systematic changes in attitude without any loss in the essential continuity of the beds as a whole, as shown in **5**.

The simple view regards folds as occurring in the form of *anticlines* and *synclines*.

As the name suggests, anticlines are arch-like structures, although often with straight fold-limbs, which generally dip away from one another in opposite directions. Synclines are the exact opposite, occurring as down-folds wherever the fold-limbs dip towards one another. Under normal circumstances, the rocks lying in the core of an anticline are stratigraphically the oldest, while synclines have fold-cores of younger rocks. However, where these stratigraphic relationships cannot be established in areas of structural complexity, or if they have been reversed as the result of earth-movements, the corresponding folds are best known as upward-closing *antiforms* and downward-closing *synforms* (see **219**).

Earth-movements can also cause rocks to fracture, forming joints and faults. *Jointing* is seen in virtually every exposure, wherever a series of closely-spaced fractures is present, lacking any visible displacement (see **113**). *Faulting* occurs wherever the original continuity shown by sedimentary or any other rocks has been broken by movements along what is known as a *fault-plane* (see **122**). A typical fault is shown in **6**, where it results in an abrupt contact between metamorphic rocks forming the rather featureless and low-lying ground to the left, and well-bedded limestones of Mesozoic age, forming the summits of the hills.

Figure 4 *Bedding-plane exposed by erosion to show how sedimentary rocks are tilted away from the horizontal as a result of earth-movements. The dip and strike of sedimentary beds can be measured in the field wherever such a bedding-plane is exposed, as described in the text. South Head (ND 376497), Caithness, Scotland.*

Figure 5 *Folding of sedimentary rocks as evidence of earth-movements. Each fold consists of a pair of* fold-limbs, *dipping at a fairly constant angle to one another, separated by a* fold-hinge, *across which the bedding changes rapidly in attitude as it is traced from one fold-limb to the next. Widemouth Bay (SS 198035), Cornwall, England. (Height of section c 5m)*

Figure 6 *Faulting as another response to earth-movements, causing the juxtaposition of rocks not originally in contact with one across a* fault-plane. *The line of such a fault-plane forms a well-marked feature of the landscape in Crete, across which metamorphic rocks occur in faulted contact with well-bedded Mesozoic limestones. Lastros, Crete.*

13

SEDIMENTARY STRUCTURES AND THEIR CLASSIFICATION

Sedimentary structures broadly include all the structural features which are formed in sedimentary rocks as a result of the various processes leading to their deposition and subsequent consolidation. An understanding of these structures is important to the geologist in the field because they can be used to determine the processes involved in the deposition of sedimentary rocks, the environments and conditions of deposition, and the directions of the *palaeocurrents* (Greek: *palaios*, old) which originally deposited the sediment. However, they are of prime interest to the structural geologist as *indices of stratigraphic order*. Earth-movements can obviously affect the bedding of sedimentary rocks in such a way that it becomes increasingly tilted away from the horizontal until it eventually reaches the vertical, beyond which it becomes overturned. Such inversions of the bedding typically occur in areas of structural complexity, where folding and faulting have affected the rocks to a very considerable extent. Structural geologists then place great reliance on the use of sedimentary structures to determine the original tops and bottoms of the beds. Any sedimentary structure to be used in this way needs to show a degree of asymmetry so that it does not have the same appearance when upside-down. It then acts as an index of stratigraphic order, allowing the way-up or *"younging"* of the sedimentary rocks to be determined. Indeed, this is the only way that a full understanding of the geological structure can be achieved in such areas of structural complexity.

Classification of Sedimentary Structures

Depositional Structures   First, there are sedimentary structures formed as a direct result of depositional processes. Apart from bedding itself, which obviously belongs to this category, these structures are typically found as internal features within sedimentary beds, or occur on the bedding-planes forming the upper surfaces of these beds. Although they are mostly physical in origin, being produced by the currents as sediment is deposited, some structures such as stromatolites (see **33**) are formed as a result of organic activity.

Since the great majority of sedimentary rocks are laid down under water, depositional structures are mostly a result of hydraulic processes. Such processes are found to act in a wide variety of depositional environments, so that depositional structures by themselves cannot be expected to tell us much about the nature of the sedimentary environment. This is confirmed by the fact that many depositional structures are found in rocks laid down under many different environments, so that they are not diagnostic of any particular environment. Indeed, the processes of sedimentary deposition are much the same in deserts, where the sedimentary load is carried by the wind and air is the "hydraulic" medium for sedimentary transport, as under water. Only the transport of sedimentary material by glaciers involves so physically distinct a process, that the resulting deposits can reasonably be taken as diagnostic of a particular environment of deposition.

Erosional Structures Next to be considered are the sedimentary structures formed by erosion of the underlying sediment, prior to the deposition of the overlying bed. Erosion usually occurs as a response to currents of air or water, flowing across a surface of fine-grained sediment, which then deposit coarser-grained sediment on top. This sediment fills in any irregularities that were originally developed in the underlying surface, preserving the form of these erosional structures as *casts* on the bottom of the overlying bed. Where such a cast is exposed, it forms a mirror image of the original structure. *Sole structures* of this sort may be formed directly by the current or by various "tools" being dragged, rolled or bounced across the underlying surface by the force of the current.

Post-depositional Structures Finally, there are numerous structures produced by post-depositional processes acting on sedimentary rocks, prior to their final consolidation. These structures are sometimes known as *penecontemporaneous* since they may well form immediately after deposition, while the sediment remains soft and incoherent. Other structures of this type may be associated with the actual processes of consolidation.

Various types of post-depositional structures can be recognized. Firstly, there are the sedimentary structures produced by physical disturbance of unconsolidated sediment. Many structures of this type occur in response to conditions of gravitational instability. For example, where any sediment is deposited on top of less dense material, it is possible for differential subsidence to take place. Likewise, if sediment is deposited on a slope, down-slope *slumping* and *sliding* can occur. Unconsolidated sediment can also be disrupted after its deposition in a variety of other ways, perhaps as a result of earthquake shocks, or changes in physical condition. The effects of *dewatering* as sediment undergoes compaction and consolidation must also be important, since any water occupying the pore-spaces between the sedimentary grains must first be expelled.

Secondly, the activity of living organisms moving across and through newly deposited sediment commonly results in its physical disturbance, giving rise to sedimentary structures known as *trace fossils*. Finally, there is a variety of chemical processes which occur in sedimentary rocks as they undergo consolidation as the result of diagenesis. The redistribution and concentration of chemical material within the rock then results in *diagenetic structures*.

Depositional Structures

Parting Lineations

This depositional structure is commonly associated with sandstones, deposited from rapidly flowing currents, which show a *sedimentary lamination* up to 1cm in thickness, parallel to the upper and lower surfaces of the bed itself. **7** shows such a *parting lineation*, exposed on a series of closely-spaced bedding-planes in a sandstone. The ridges defining such a lineation are usually only a few millimetres apart, while they may persist for several centimetres or more along their length. They form parallel to the flow-direction of the current which deposited the sediment, although the actual sense of flow cannot be determined.

Ripple-marks

Commonly seen at the present day, ripple-marks are formed by the movement of water over a surface of loose sediment, often fine sand or coarse silt, throwing it up into a series of low ridges, separated from one another by furrows. Such morphological features are generally known as *bed-forms*. The ridges are usually spaced up to 50 or 60cm apart, often much less, with a difference in height between trough and crest measuring up to 3cm. There are two distinct ways in which this structure can be formed.

Wave-formed ripples are produced by the action of waves on standing water. Each wave in passing generates eddies which oscillate back and forth across the bottom. This results in the formation of symmetrical ripple-marks, as seen in cross-section in **8**. They typically have broad and rather rounded troughs, separated from one another by much narrower crests, which are often somewhat cuspate in shape. This type of ripple-mark can be used as an index of stratigraphic younging under suitable circumstances. It is sometimes known as an oscillation ripple-mark, particularly if the internal laminae adopt a chevron-like pattern on either side of the ripple crests.

Current-formed ripples are formed by the eddying action of a current flowing across a sedimentary surface. As shown in **9**, the ripple-marks formed in this way are asymmetrical in profile, each with a steep *lee-side*, facing down-current, and a gentle *stoss-side*, facing in the opposite direction. This allows the direction of flow to be determined for the current which formed the ripple-marks. Although there is a difference in shape between the crests and troughs of current-formed ripples, it is rarely so marked that it can be used as a reliable index of stratigraphic younging.

Figure 7 *Parting lineation trending from top left to bottom right across the bedding-planes of a laminated sandstone. It forms a series of closely-spaced ridges of very low relief, best seen under slanting light. Often, the rock splits apart along a number of small steps, parallel to this direction. St Monance (NO 534018), Fife, Scotland. (Field of view c 60cm)*

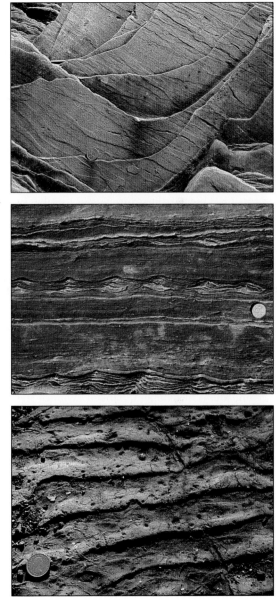

Figure 8 *Symmetrical ripple-marks affecting silty layers, showing their characteristic form as seen in cross-section. The crests of the ripples migrated to the right during the course of deposition, so that the internal laminae making up these ripples are also inclined predominantly in the same direction. Scremerston (NU 018502), Northumberland, England.*

Figure 9 *Asymmetrical ripple-marks on a bedding-plane, showing that each ripple has one side consistently steeper than the other, facing downstream in the direction of flow of the current. The present example therefore shows that the current flowed from top left to bottom right across the bedding-plane as it is now exposed. Brunton Bank (NY 018502), Northumberland, England.*

Cross-bedding

Cross-bedding occurs wherever the internal stratification within a bed is inclined, more or less regularly, at an oblique angle to the bedding-planes defining the bed itself (see 10). It is a common type of sedimentary structure, often present in sandstones and other detrital sediments but also found in limestones and dolomites. Terms such as current-bedding, false-bedding and inclined-bedding, formerly applied to cross-bedding, are now obsolete. Cross-bedding mostly forms in response to the migration of sand-dunes, sand-waves and ripples in the direction of current flow, downstream or downwind, under conditions where sedimentary material is actually being deposited at the same time. Strictly speaking, cross-bedding should be restricted to cases where the inclined layers within the bed are thicker than a centimetre. If this is not the case, the structure really ought to be termed *cross-lamination*, while the two types of sedimentary structure would together be considered as examples of *cross-stratification*.

Sand-dunes and Sand-waves   These morphological terms are applied to bed-forms similar to ripples but developed on a much larger scale. *Dunes* are closely-spaced bed-forms with crests up to 5m apart, showing a considerable degree of relief between trough and crest, and shaped in response to the flow of relatively fast currents. They frequently have highly sinuous crests, while the intervening troughs are often accentuated by the scouring action of the currents. The shape of sand-dunes is therefore highly complex in three dimensions. Typically, the lee-sides of sand-dunes form scour-hollows, facing down-current. *Sand-waves* are much more uniform in their morphology. They consist of widely-spaced and somewhat asymmetrical ridges with straight or slightly sinuous crests, often more than 5m apart, formed by the flow of more sluggish currents. Typically, they have the form of *sand-bars*, which occur as flat-topped accumulations of sediment, flanked by much steeper lee-sides, facing down-current. There appears to be a complete transition in morphology between sand-dunes and sand-waves. Since they mostly occur under some depth of water, present-day examples of sand-waves are rarely seen. However, ripple-marks show a similar range in morphology, on a smaller scale.

Mode of Origin   The crests of sand-dunes, sand-waves and ripples all migrate downstream in the same direction as the current (see Drawing 1, p.18). This does not simply occur as a result of sedimentary particles being transported across a whole series of bed-forms without any pause. Instead, sedimentary particles are eroded from the stoss-sides of each bed-form, transported across its crest, and deposited on its lee-side, to create the *foreset beds*. These beds gradually encroach over the erosion surface, formed by the stoss-side of the next bed-form lying immediately down-current from the first. The resulting structure will be preserved provided that deposition exceeds erosion overall. However, unless the supply of sediment greatly exceeds the power of the current for its transport, only the foreset beds are preserved in this way. This gives rise to a series of internal bedding-planes, inclined at angles up to 30° or thereabouts in the direction of flow, which is characteristic of cross-bedding.

Deltaic Cross-bedding This type of cross-bedding can be identified by the presence of a single cross-bedded unit, up to several metres or even more in thickness, which can often be traced laterally for a considerable distance. How such a unit was formed by a delta-front advancing into standing water has been taken as the traditional model for the development of cross-bedding. However, it is now realized that this mode of origin is very rare although it is an explanation which is still often found in the elementary text-books. It results in a series of foreset beds, dipping at angles between 10° and 25°, which usually consist of sand. They overlie deposits of silt and mud, into which they pass down-slope, so forming the *bottom-set beds*, which were deposited in deeper water in front of the delta. *Top-set beds* formed by gravels, sands and silts are commonly found lying on top of the cross-bedded unit. Deltaic cross-bedding developed on such a scale is only a rare feature of the geological record. This occurs because it is only likely to form in a lacustrine (Latin: *lacus*, a lake) environment, which is itself not very common in the geological record. Larger deltas are of course quite a common feature of sedimentary sequences.

Environmental Significance Although the different types of cross-bedding can be used to distinguish between a variety of sedimentary environments, this is well beyond the scope of the present book. However, apart from cross-bedding developed in silty rocks, which can be formed in many different environments, it is a structure that is characteristic of deposition in shallow water. Typical environments include braided and meandering rivers, lakes and lagoons, coasts and shore-lines, and the shallow seas surrounding the continents. Apart from any environmental significance that cross-bedding may possess, it can also be used to determine the direction of flow for the currents that actually deposited the sediment. This corresponds to the down-dip direction shown by the foreset beds, or its mean value, corrected as necessary for any subsequent tilting of the bedding as a whole.

Use in Structural Geology The prime interest of cross-bedding for the structural geologist lies in its use as an index of stratigraphic younging, particularly where it is tangential in character (see **12**). Not only is cross-bedding commonly found in sedimentary rocks which have been profoundly affected by earth-movements, but it also tends to occur in rocks which often resist the effects of deformation and metamorphism. This commonly results in the complete obliteration of other types of sedimentary structure, which could otherwise be used as indices of stratigraphic order. Even so, recrystallization and the formation of quartz veins in particular often make cross-bedding increasingly difficult to recognize, while extreme deformation may reduce the angle made by the foreset beds with the external bedding-planes to zero, so that it cannot be recongized at all in such rocks.

Types of Cross-bedding

Cross-bedding is best seen in vertical cross-sections at right angles to the bedding, but orientated parallel to the direction of dip of the foreset beds. **10** shows its characteristic features. Each cross-bedded layer is known as a *set*, while the whole series of such units occurring within a single bed is called a *coset*. The currents depositing the sediment in the present case evidently flowed from left to right, judging by the orientation of the foreset beds. Occasionally, the foreset beds in adjacent cross-bedded units dip in diametrically-opposed directions, giving rise to what is known as *herring-bone cross-bedding*. Such changes in dip of the foreset beds must result from reversals in flow of the palaeocurrents depositing the sediment, which are most likely to occur in response to the ebb and flow of tidal currents.

Angular and tangential cross-bedding

There are two distinct types of cross-bedding, to judge by the form of the foreset beds as seen in cross-section. They are best distinguished as angular cross-bedding and tangential cross-bedding. **11** is an example of *angular cross-bedding*. This morphology is formed as a result of sedimentary grains avalanching down the lee-side of the bed-form. It should be stressed that this form of cross-bedding cannot be used as an index of stratigraphic order, as may be appreciated by turning the page upside down. The foreset beds do not have a concave-upwards form, typical of tangential cross-bedding, while the top and bottom contacts of the cross-bedded unit cannot easily be distinguished from one another.

12 is an example of *tangential cross-bedding*. It can be seen that the foreset beds vary in dip in such a way that they make an asymptotic contact with the underlying rocks. This means that the foreset beds have a shape which is distinctively concave-upwards, while there is a marked contrast between the tangential contact at the base of the unit and the erosion surface or "cut-off", which cuts across the foreset beds at its top. It is these features which make this type of cross-bedding so useful an index of stratigraphic order. They are formed wherever the avalanching of sedimentary grains down the foreset beds is augmented to a greater or lesser extent by the eddying action of the current in the intervening troughs.

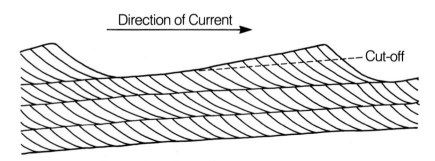

Direction of Current

Cut-off

Drawing 1 *How cross-bedding is formed by the down-current migration of sand-bars.*

Figure 10 *Cross-bedding in a Carboniferous sandstone, showing its typical appearance in the field. Overall, the bedding is horizontal, forming a series of cross-bedded layers within which the foreset beds consistently dip at angles up to 25° towards the right. Bowden Doors (NU 071326), Northumberland, England.*

Figure 11 *Angular cross-bedding in a sandstone, showing how the foreset beds dip at a uniform angle throughout the cross-bedded unit. Note how these foreset beds make an angular contact with the underlying bedding-plane. Corbridge (NZ 000654), Northumberland, England.*

Figure 12 *Tangential cross-bedding in a quartzite forming a wedge-shaped unit overlain by evenly-bedded rocks and resting on another cross-bedded unit underneath. This form of cross-bedding provides a very reliable index of stratigraphic order, unlike angular cross-bedding, which cannot be used in this way. Caolsnacon (NN 142607), Argyll, Scotland.*

Tabular and Trough Cross-bedding

The morphology of cross-bedding in three dimensions depends on whether straight-crested sand-waves or crescentic sand-dunes were involved in their formation. The down-current migration of straight-crested sand-waves results in *tabular cross-bedding*, as shown in **13**. The tabular nature of such cross-bedded units means that they can often be traced for considerable distances before they eventually disappear, often beyond the confines of a single exposure.

By way of contrast, crescentic sand-dunes in migrating down-current generate *trough cross-bedding*. The foreset beds then change in dip and strike to form troughs or, more accurately, scoop-like hollows facing down-current. The upper and lower boundaries to these units are rarely parallel to one another, so that the individual units are seen to be lenticular or wedge-shaped in most cross-sections. The foreset beds within these units generally make tangential contacts with the underlying rocks.

These two types of cross-bedding are best distinguished from one another where their three-dimensional form can be determined. This is most easily done wherever the erosion surface forming the upper contact of a cross-bedded unit is exposed. The underlying foreset beds will then intersect this surface in a series of traces.

Tabular cross-bedding can be identified by the relatively straight traces developed by the foreset beds on this surface, whereas the foreset beds in trough cross-bedding give rise to a series of crescent-like traces on this surface, as shown in **14**, with a concave sense of curvature.

These foreset traces can also be used to determine palaeocurrent directions. This is done in the case of tabular cross-bedding by taking the direction at right angles to these foreset traces, corresponding to the dip of the foreset beds. The observations need to be corrected for any subsequent tilting of the bedding. Dealing with trough cross-bedding is more complicated, since only the trend of the trough-like traces shown by the foreset beds on this erosion surface can be used directly to determine the palaeocurrent direction.

Trough cross-bedding can also be identified where a structure known as *festoon cross-bedding* is seen in vertical cross-sections, as shown in **15**. This term should not be used to imply that it is morphologically distinct from trough cross-bedding, since this is not the case. It is simply trough cross-bedding as seen in section cutting vertically across the troughs, rather than along their lengths (as **12**) or horizontally in plan-view (as **14**).

Figure 13 *Tabular cross-bedding in a sandstone, in which each cross-bedded unit has a tabular form, with its upper and lower surfaces effectively parallel to one another. The foreset beds often have much the same dip and strike throughout each unit, while they usually make an angular contact with the underlying rocks. Corbridge (NZ 000654), Northumberland, England.*

Figure 14 *Trough cross-bedding, seen in horizontal cross-section, where the foreset beds give rise to strongly curved traces on the erosion surface overlying the cross-bedded unit. These traces define a series of crescentic shapes, facing down-current away from the observer. Longhoughton (NU 263157), Northumberland, England.*

Figure 15 *Festoon cross-bedding, formed wherever the foreset beds in trough cross-bedding are seen in vertical section at right angles to the trend of the individual troughs. These foreset beds then occur as festoon-like features within each cross-bedded unit. Whitley Bay (NZ 362722). Tyne and Wear, England.*

Ripple-drift Bedding This type of wavy cross-bedding is produced on a small scale in rapidly-deposited sediment wherever ripples are formed by the currents supplying the sediment itself, as shown in **16**. Strictly speaking, it should be termed cross-lamination if the foreset beds are less than a centimetre in thickness. Often, ripple marks are preserved on the bedding-planes, particularly where they are formed by seams of shaly material, interspersed with more silty material. Although these ripples may be straight-crested, or only somewhat sinuous in plan, they frequently have crescentic forms, facing up-current (linguoid ripples) or down-current (lunate ripples). These terms are derived from the Latin (*lingua*, a tongue; *luna*, a moon). Lunate ripples in particular can result in the formation of what is known as *rib-and-furrow* on the bedding-planes. This consists of a series of shallow ridges and troughs, which are formed by the foreset beds being arranged in a scallop-like fashion, facing down-current. It occurs as a small-scale replica of trough cross-bedding.

If the current is more heavily charged with sediment than usual, or if the waning force of the current can no longer carry so much of the sediment, deposition comes to outweigh the effects of erosion, particularly towards the top of a bed. This is first marked by the foreset beds becoming somewhat sigmoidal in shape, so that the bedding can be traced for greater and greater distances across the crests and troughs of individual ripples, as shown in **17**.

This effect is accentuated wherever more sediment comes to be deposited on the stoss-sides of the ripples, as well as their lee-sides, owing to the currents losing their power to carry along the sediment. Each ripple tends to migrate down-current over the backs of those lying farther downstream, to form what is known as *climbing ripples*, as shown in **18**. Note how the bedding can be traced without interruption across these ripples. It is then found that the crests and lee-sides are formed by silty sediment, which is light in colour, while more shaly material occurs in the intervening troughs, where it forms the darker rock.

Various terms are applied to the lithological features formed by the migration of ripples. *Wavy bedding* is found wherever silty layers showing the effects of ripple-drift bedding are separated from one another by thin but continuous seams of muddy rock. *Flaser bedding* (German: *flaser*, streaks or lenticles) occurs wherever muddy material forms discontinuous streaks or arcuate lenses, lying concave-upwards in the troughs formed as a result of ripple-drift bedding. *Lenticular bedding* is present wherever silty material occurs in the form of isolated ripples, internally cross-bedded and still retaining their external form, which are more or less surrounded by muddy sediment.

Figure 16 *Ripple-drift bedding in silty sediment, formed by the down-current migration of ripples during the course of sedimentary deposition. Often, the profiles of individual ripples are preserved to a greater or lesser extent in this type of small-scale cross-bedding. Seaton Sluice (NZ 364757), Tyne and Wear, England.*

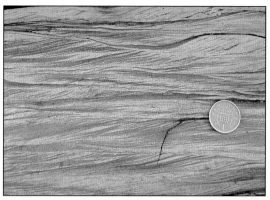

Figure 17 *Ripple-drift bedding in which the stoss-sides of individual ripples tend to be preserved as well as their lee-sides. Note the general lack of scoured surfaces in this example. Seaton Sluice (NZ 364757), Tyne and Wear, England.*

Figure 18 *Climbing ripples formed wherever the continuous deposition of sediment preserves the form of current ripples as they migrate slowly downstream. A lithological banding is inclined in the opposite direction to the dip of the foreset beds, mimicking the form of cross-bedding on a larger scale. Bude (SS 201073), Cornwall, England.*

Dune-bedding Aeolian cross-bedding (Latin: *Aeolus*, god of the winds) can be distinguished from all other forms of cross-bedding, which are produced by the effects of flowing water, since it results from the action of the wind. The foreset beds then represent the lee-sides of aeolian sand-dunes. This type of cross-bedding typically occurs in the form of much larger units than usual, several metres or even much more in height, while the foreset beds commonly dip more steeply at angles in excess of 30°. **19** shows that the whole of the deposit usually consists of cross-bedded units, giving rise to a structure best known as *dune-bedding*. The highly irregular nature of dune-bedding, which is another characteristic feature, arises partly from the presence of trough cross-bedding on a large scale, coupled with the considerable variations in attitude of the cross-bedded units, exaggerated by the steep dips of the foreset beds. It is often difficult, if not impossible, to find the overall dip and strike of such a deposit.

Topset Cross-bedding This term is best applied to a form of cross-bedding in which the topset beds are preserved as well as the foresets. An example is shown in **20**. The dark layer at the bottom is a shaly bed, associated with silty material showing ripple-drift bedding. A small-scale delta has built itself out across this layer, depositing the rather massive sandstone on which the coin rests. Although the foreset beds within this sandstone are not very clear, they do dip towards the right. They pass up-dip towards the left into a series of shales together with more sandy layers, which represent the topset beds, overlying this delta. They are covered in their turn by more silty sediments, showing good examples of ripple-drift bedding. Note that the foreset beds, in passing up-dip into the topset beds, are concave-downwards in shape, while they appear to rest with an angular contact on the underlying rocks.

Overturned Cross-bedding Although it is, strictly speaking, a post-depositional structure, the overturning of the foreset beds in cross-bedded units may be considered at this point. **21** shows a typical example. Although the overturning inevitably takes place in a direction corresponding to the dip of the forest beds, it is not entirely clear that it occurs as a result of the frictional drag exerted by the currents which deposited the cross-bedded unit in the first place. There is a possibility that such a structure may be formed as a result of dewatering.

Figure **19** *Dune-bedding in aeolian sandstone, showing how this type of cross-bedding typically occurs in large and wedge-shaped sets, representing the lee-sides of aeolian sand-dunes. Corrie (NS 026428), Isle of Arran, Scotland.*

Figure **20** *Topset cross-bedding formed wherever the topset beds have not been removed as the result of subsequent erosion. Such a structure, if identified by mistake as the more usual type of cross-bedding, would suggest that the beds were upside down, so that care must be taken in its interpretation. Pittenweem (NO 543032), Fife, Scotland.*

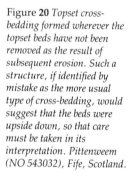
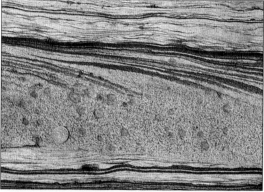

Figure **21** *Overturned cross-bedding in which the foreset beds first become steeper than usual, and then overturned, as they are traced towards the top of a cross-bedded unit. This structure is sometimes found in association with convolute bedding (see **55**). Bowden Doors (NU 071326), Northumberland, England. (Field of view c 2.5m)*

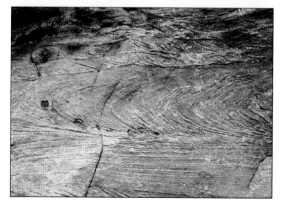

Graded Bedding

Graded bedding typically occurs wherever the sedimentary particles within a single bed become finer in grain-size as they are traced upwards throughout its entire thickness. A typical example of such a graded bed is shown in **22**. It rests with a sharp contact on the underlying sediment, which is much finer-grained, showing the development of ripple-drift bedding towards its very top. The sedimentary particles visible to the naked eye clearly become smaller, and much less numerous, towards the top of the graded unit, so forming a pebbly layer which passes upwards over a short distance into a much finer-grained sandstone.

The sharp bases typically shown by graded beds are the result of internal grading within each bed. This results in the juxtaposition of coarser-grained sediment on top of much finer-grained material, which forms the top of the graded bed lying immediately underneath, as shown in **23**. Note how erosion and scouring of the underlying sediment have accentuated the sharpness of this contact.

Use in Structural Geology Graded bedding is perhaps the most useful index of stratigraphic order available to the structural geologist. This is partly because it is frequently found in sedimentary rocks, which have subsequently been affected by deformation and regional metamorphism. However, it also tends to resist the obliterating effects of these processes better than most other sedimentary structures, particularly where it is found in coarse-grained sediments. By way of contrast, cross-bedding often becomes increasingly difficult to recognize in these circumstances.

Graded bedding can be recognized from any differences in grain-size within a sedimentary bed, which are visible to the naked eye, augmented as necessary by the use of a hand-lens. It is often a good idea to remove *small* chips from the base of the bed, which can then be compared side-by-side with its top. In finer-grained sediments such as siltstones, it is often necessary to rely simply on texture and colour as indicators of grain-size (see **220**). Although silty material is often lighter in colour than shale, this is not always the case, so care needs to be taken.

Reverse grading has been described in the literature, although it mostly appears to be restricted to volcanic deposits. Metamorphism (*q.v.*) can also cause reversals in grain-size within a sedimentary bed, as shown in **24**. Often, metamorphic minerals such as garnet and andalusite occur preferentially towards the top of a graded bed, where the rock has a chemical composition most appropriate to their growth. This produces a mimetic change in grain-size which is the exact reverse of that originally found so that the present rock shows the rock itself to be stratigraphically inverted.

Figure 22 *Graded bedding formed wherever there is a gradual decrease in grain-size within a sedimentary bed, passing upwards from coarser-grained sediment at its base into finer-grained material towards the top. Knock Head (NJ 661659), Banffshire, Scotland.*

Figure 23 *Erosional base to a graded bed, marking an abrupt change from the much finer-grained sediment underneath. Even if such a bed cannot be seen to be graded, it is often found that its undersurface is much rougher to the touch than the bedding-plane forming its top. New Quay (SN 385604), Dyfed, Wales.*

Figure 24 *Reverse grading in a metamorphic aureole. The prominent crystals are garnets, weathering proud of the surface, which form a few large crystals in the more shaly part of the bed, while they occur as smaller but more numerous crystals where the rock grades into siltstone. Spar Craigs (NJ 933197), Aberdeenshire, Scotland. (Field of view c 2cm)*

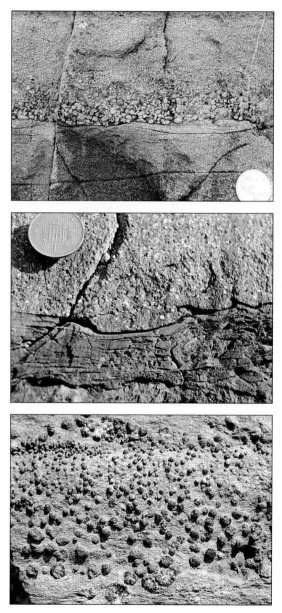

Graded Bedding and Turbidite Deposition

Graded bedding results from a gradual waning in the force of the current which deposits the sedimentary bed. As the current declines, it is the heaviest, and therefore usually the largest particles, which settle out first. They are followed by the deposition of sedimentary particles which become lighter, and therefore smaller, as deposition proceeds. The largest particles then occur towards the base of the bed, while smaller and smaller particles are found towards its top. This mode of deposition is typical of what are known as *turbidity currents*, although it is not restricted by any means to such currents.

As the name suggests, turbidity currents are turbulent suspensions of sand, silt and mud, which can flow down even the slightest slopes under water by virtue of their excess density. The initial stages in the flow of such a current are marked by increasing velocities of flow, which then reach a maximum value, before they eventually start to decline in the waning stages of the current. The early stages in the flow of the current before deposition starts are marked by the formation of *sole structures* on the base of the bed which is subsequently deposited on top. These structures are formed by the erosional scouring of the muddy floor over which the turbidity current passes with its load of coarser-grained sediment. They are described in the next section, dealing with erosional structures. Such structures are, however, typical of *turbidites*, which is the name given to the sedimentary deposits left by turbidity currents.

Turbidite Sequences   Although turbidity currents can form in many different environments, the great majority of turbidites preserved in the geological record are deep-water deposits. They form sedimentary sequences up to a few thousand metres thick, or sometimes even more, consisting of a monotonous alternation of sedimentary beds, differing from one another only in grain-size. The finer-grained beds are shales, which were evidently laid down in quiet conditions. Any fossils that they contain are mostly deep-water forms, or the remains of organisms that lived near the surface of the open sea. These shales are interbedded with coarser-grained sediments, forming relatively thin beds which nevertheless show remarkable lateral persistence. These beds are usually sandstones, often with the composition of greywacke. They vary in thickness from a few centimetres up to several metres. Any fossils occurring within these beds are typically shallow-water forms which have been transported from elsewhere by the turbidity current that deposited the sandstone itself. The thicker the bed, the coarser the rock, at least as a general rule. Coarse-grained sandstones may pass into fine-grained conglomerates, while fine-grained sandstones often merge into siltstones at the other end of the lithological spectrum. Occasionally, the coarse-grained components of a turbidite sequence are formed by clastic limestones, showing all the features otherwise typical of turbidite deposition.

Silty layers in a slate, showing graded bedding and affected by folds with an axial-planar cleavage. Islay, Scotland.

31

Bouma Sequences Graded bedding in turbidites is often accompanied by other sedimentary structures arranged in a consistent manner within each bed to form what has been termed a *Bouma sequence*. A typical example is shown in **25**. The very base of a graded bed may occasionally be relatively fine-grained, with the coarsest material appearing only after a few centimetres. Usually, the coarsest sediment occurs immediately at the very base of the unit. This may be marked by a thin layer showing a concentration of coarse sand, or even small pebbles (see **22**). Such a coarse layer then passes upwards into finer-grained sediment, so forming the graded part of the turbidite bed. This part of the Bouma sequence is usually massive in character, without any internal lamination (see **25**) forming the bulk of the bed. However, coarser-grained beds may show streaks and stringers of pebbly material at this level, possibly defining a rudimentary form of cross-bedding in some cases. Large fragments of mudstone, torn up from the sea-floor as the turbidity current passed over its muddy substratum, are often incorporated into this division, to form what is termed a *mud-clast breccia*.

As shown in **25**, the graded division in a Bouma sequence passes up in its turn into finer-grained sediment which typically shows parallel lamination on a small scale. This unit is commonly much thinner than the underlying part of the graded bed. It is usually composed of fine sand, passing upwards into very fine sand or coarse silt, as the parallel lamination gives way to ripple-drift bedding in the same direction towards the top of the bed. Climbing ripples (see **18**) are often found in the uppermost part of this division. It may be succeeded by alternations of fine silt and silty mud, arranged in the form of thinly-laminated beds, which pass upwards in their turn into muddy sediment, lacking much in the way of obvious bedding.

Proximal and Distal Turbidites Apart from very thick units of pebbly sandstone, most turbidites can be traced along their strike for considerable distances without varying very much in thickness or lithology. However, there is a very gradual change in their character, which corresponds to a change in the nature of the sediment deposited along the path of a turbidity current. In particular, the various divisions within a graded bed tend to disappear laterally as the bed is traced in a down-current direction, starting with the lowermost elements of the Bouma sequence. For example, *proximal turbidites*, deposited nearest to the sediment source, tend to preserve the complete Bouma sequence. **26** shows the typical appearance of proximal turbidites. The graded division of each bed is much thicker than the other elements in the Bouma sequence, which may be missing altogether, particularly towards the top of the bed. Any shale horizons interbedded with proximal turbidites are relatively thin, and may be completely lacking.

Such sediments pass into *distal turbidites*, deposited at a much greater distance from the sediment source in a down-current direction. There is a gradual decrease in thickness, accompanied by a corresponding reduction in the overall grain-size of the turbidite beds, as they are traced in this direction, since all the coarser-grained sediment has already been deposited farther up-current. These changes are also marked by the progressive disappearance of the lower elements in the Bouma sequence, starting with the graded division, so that only the uppermost elements are eventually preserved. There is also a gradual increase in the relative proportion of shale, which separates the turbidite beds from one another, as shown in **27**. All these beds ultimately pass into shales, deposited well beyond the reach of any turbidity current.

Figure 25 *Bouma sequence showing how a graded bed often passes upwards into a laminated portion, itself overlain by even finer-grained sediment with ripple-drift bedding. A shaly horizon then intervenes, often marked by silty laminations in its lower part, before the abrupt base of the next bed is encountered. Roman Steps (SH 652304), Gwynedd, Wales.*

Figure 26 *Proximal turbidites of Silurian age, showing the typical appearance with massive beds of coarse-grained greywacke of considerable thickness, separated from one another by only minor amounts of shale. New Quay (SN 385604), Dyfed, Wales. (Height of section c 8m)*

Figure 27 *Distal turbidites, deposited down-current from the proximal turbidites shown in the previous photograph. They occur as thin beds of comparatively fine-grained sediment, up to a few centimetres thick, interbedded with considerable amounts of shale. Clarach (SN 587836), Dyfed, Wales.*

Reef Limestones

Although most limestones occur normally as bedded sequences, *reef limestones* are an exception. In essence, they are formed by colonial organisms which grow together to construct an organic framework known as a *reef*. Although recent examples mostly occur in the form of coral reefs, many other organisms were responsible for their construction in the geological past.

Limestone reefs in the geological record vary from small-scale structures, easily visible in a single exposure, to huge formations that measure many hundreds of kilometres along their lengths. Commonly, the small-scale examples form *patch reefs*, as shown in 28. Vertical cross-sections through a patch reef commonly show how it grows outwards from its base to reach a maximum size, and then contracts once more as it is overwhelmed by normal sediment.

Examples of reef limestones on a larger scale typically form *barrier reefs*, which are found to separate shallow-water lagoons from much deeper water lying offshore. 29 shows a splendid example of such a reef complex from Western Australia. Massive limestones lacking any obvious bedding form the core of the reef. They are flanked to the left by *fore-reef limestones*, which are well-bedded rocks formed by debris derived from erosion of the reef itself. They show what was an initial dip of 30° or thereabouts into the deeper water, so forming an apron around the reef. The horizontally-bedded limestones lying to the right represent *back-reef limestones*, deposited as shallow-water sediments in the lagoon lying behind the reef.

Geopetal Structures One way to demonstrate that fore-reef limestones were deposited with an initial dip involves the use of what are known as *geopetal structures*. The commonest type, although still of rare occurrence, is formed by sediment filling in only the bottom half of any cavities left in fossils by the decay of their soft parts. This leaves a space where mineral matter, usually calcite, can be deposited on top of the sedimentary infilling.

30 shows an example of such structures in a fore-reef limestone, where the bedding dips overall towards the left. The exposure shows numerous cross-sections through brachiopod shells, now mostly filled with calcite. However, there are several shells in which sediment fills in the lower part of these cavities. The contacts formed by this sedimentary infilling with the overlying mineral matter are all close to the horizontal, which most likely corresponds to their attitude at the time of their deposition. This means that the present dip shown by the bedding is likely to be an initial dip.

Figure **28** *Patch reef in bedded limestones, forming a discrete mass of reef limestone, interbedded with normal sediments, and lacking any obvious internal structure. Taurus Mountains, Turkey. (Field of view c 2m)*

Figure **29** *Reef complex of Devonian age in Western Australia, showing massive reef limestones in the centre, separating well-bedded fore-reef limestones with an initial dip to the left from horizontally-bedded back-reef limestones to the right. Windjana Gorge, Western Australia. (Photograph by C.T. Scrutton)*

Figure **30** *Geopetal structures forming the infillings of brachiopod shells in a fore-reef limestone. Several shells are partially filled with sediment, allowing calcite or other mineral matter to fill in the remaining space on top, so forming a very useful index of stratigraphic order. Emanual Range, Western Australia. (Photograph by C.T. Scrutton)*

Biogenic Structures

Stromatolites

A particular type of lamination is produced in fine-grained limestones by the action of blue-green algae. This lowly lifeform commonly produces an algal mat on top of the sedimentary surface, which then traps carbonate and other detrital particles. The laminations are the result of slight variations in the sedimentary influx, since the algae can grow up through any thin layers of sedimentary particles deposited on top of the mat. **31** shows the typical appearance of such a sedimentary lamination.

Algal laminations formed in this way often show a variety of different growthforms, which are then known as *stromatolites*. Typically, they occur in the form of mounds, domes and columns, all packed closely together. **32** shows recent examples of such stromatolites from a tidal flat in Western Australia, while **33** shows the laminated structure typical of stromatolites in cross-section.

Most stromatolites are found as such dome-like masses, often lying in contact with one another, so that the internal laminations have a cuspate form. It is this feature which allows stromatolites to be used as an index of stratigraphic order, particularly useful where they occur in Precambrian limestones and dolomites, dating back to 3000 million years ago. The cusps all point downwards into the stratigraphically older beds, while the intervening stromatolitic masses are always defined by sedimentary laminations showing upwardly-convex shapes, whatever their detailed form.

Figure **31** *Cryptalgal laminations in a fine-grained limestone, formed by the trapping of sedimentary particles by mats of blue-green algae, growing over the surface of the sediment. Vrises, Crete.*

Figure **32** *Stromatolites growing at the present day on a tidal flat, showing their characteristic form. Shark Bay, Western Australia (Photograph by C.T. Scrutton)*

Figure **33** *Stromatolites as seen in cross-section, showing their characteristic form. This example must have formed a low mound, above the level of the surrounding sediment, which gradually became larger in size as it grew upwards and outwards during the course of deposition. Bonnahaven (NR 423729), Islay, Scotland. (Field of view c 20cm)*

Oncolites A particular type of stromatolitic growth-form produces what are known as *oncolites*, or algal balls. They occur in the form of spheroidal masses, unattached to their sub-stratum, and often showing a concentric lamination when seen in cross-section. **34** shows a fore-reef limestone packed with oncolites, which have been swept off the top of a drowned reef into deeper water. The bedding dips quite steeply towards the right. Note how the oncolites forming the uppermost layer of this deposit are capped by vertical growths. These caps evidently formed on top of each oncolite once it had reached its present position, showing that these fore-reef limestones were deposited with an initial dip close to 40°.

Fossil Growth Positions

Another biogenic structure which can be used as an index of stratigraphic younging is provided by fossils, wherever they are preserved *in situ* in their positions of growth, under relatively tranquil conditions. Although much palaeontological knowledge is often required for their interpretation, the positions of growth for some fossil species can easily be identified in the field. For example, **35** shows a colonial coral in its growth position, with the individual corallites branching upwards from their base. Likewise, brachiopods are often found in their positions of growth, each with its larger valve embedded in a concave-upwards position in the underlying sub-stratum.

However, considerable care is needed in the interpretation of this evidence. Colonial corals can be turned upside-down through the action of waves and currents, while the shells of brachiopods, broken apart from one another, often come to rest, facing concave-downwards on their sub-stratum, in a position of dynamic equilibrium. For example, **36** shows productid brachiopods in what appears to be a position of growth, with their larger valves facing in a concave-upwards direction. However, there is a layer, seen towards the top of the photograph, where these valves have apparently been turned upside-down, so that they now mostly face in a concave-downwards direction. That this is not their normal position of growth is perhaps suggested by the rather varied attitudes that are adopted which are shown by these brachiopod valves where they occur within this layer.

Figure 34 *Oncolites in a fore-reef limestone, capped by vertical growths. This is another example of a geopetal structure, this time defining the vertical rather than the horizontal at the time of deposition. Canning Basin, Western Australia. (Photograph by C.T. Scrutton)*

Figure 35 *Colonial coral* (Thamnophyllum germanicum) *preserved in its growth position within the Middle Devonian rocks of the Daddyhole Limestone Formation, Torquay (SX 922628), Devon, England. (Photograph by C.T. Scrutton)*

Figure 36 *Productid brachiopods as seen in cross-section, illustrating the difficulties in deciding whether or not fossils are preserved in positions of growth. Loose block, Wiseman's Bridge (SN 147061), Dyfed, Wales*

Erosional Structures

Sedimentary structures formed by the erosion of an underlying sub-stratum fall into two distinct categories, depending not only on how they were formed but also on the manner in which they are seen to occur in the field. Firstly, there are *surfaces of erosion*, which are best seen in vertical cross-sections at right angles to the bedding. They are often associated with the development of *channels*, marking the sites of sedimentary transport by flowing water, frequently over long intervals of time. Channels are cut by rivers as they meander across their flood-plains and deltas, close to sea-level. However, they may also be produced off-shore in shallow water, largely in response to the ebb and flow of tidal currents, and in deeper water through the action of turbidity currents.

Secondly, there are the erosional structures typically preserved as *sole markings* on the bottom surfaces of sandstone beds. These are very frequently associated with the deposition of turbidites. However, storm surges in shallow seas, flash floods in semi-arid deserts, and sheet floods across alluvial plains, may all result in the formation of sole structures, very much like those found in turbidite sequences. They are best seen where the bedding is inverted, since they generally occur as casts on the bottom of an overlying bed of coarser-grained sediment, stripped bare by erosion.

Channels and Washouts

Although channels are often formed on rather a large scale, up to several kilometres across, it is only the smaller examples that are easily recognized within the confines of a single exposure. 37 shows the base of such a channel, which cuts down to the left, truncating the underlying beds as it does so. This cross-cutting relationship identifies the base of such a channel as a surface of erosion cut into older rocks, so providing a useful index of stratigraphic younging. Such features can only be developed on a relatively small scale, provided that the beds remain conformable to one another throughout the sequence as a whole. They differ in this respect from angular unconformities (see **100**).

As individual channels become smaller, they pass into erosional surfaces which are formed on a very minor scale as the result of *scour-and-fill*. An example is shown in **38**, where it can be seen that the base of what might best be termed a *washout* has a scalloped form in cross-section. The beds laid down on top of this erosion surface show a series of wave-formed ripple-marks, also in cross-section. Note how the crests of these ripples, marked by slightly coarser-grained sediment, slant upwards to the right, showing that they migrated slightly in this direction as deposition proceeded.

The shape of channels in cross-section is quite variable, although overall they are concave-upwards, as shown in **39**. Note again the contrast between the very massive sediment lying within the channel, and the well-bedded standstones showing cross-bedding on either side. However, it is just as common to find channels filled with cross-bedded sandstones, often with a conglomeratic layer at their base, formed by fragments of the underlying rocks.

Figure **37** *Channel margin defined by an abrupt contact between poorly-bedded and rather massive rocks lying within the channel itself, and the well-bedded sandstones that occur underneath. Banff (NJ 682647), Banffshire, Scotland.*

Figure **38** *Wash-out in fine-grained sediment, formed as a result of scour-and-fill. The sides of this very small-scale channel have a scalloped form, showing where the eroding currents encountered slightly more resistant horizons in cutting down into the underlying sediment. Scremerston (NU 071326), Northumberland, England.*

Figure **39** *Channel filled with massive sandstone, showing the characteristic concave-upwards form of such structures, which can itself be taken as a useful index of stratigraphic order. Bowden Doors (NU 071326), Northumberland, England. (Height of section c 3m)*

Flute-casts

Flute-casts are sole structures which are associated with turbidite beds. **40** shows the bottom of a greywacke bed, on which a whole series of flute casts are developed in a very typical fashion, while **41** is a closer view, showing their characteristic form. These structures are formed by the infilling of erosional hollows, which were scoured out by eddies as the currents flowed across a muddy sea-floor. The subsequent infilling of these hollows by coarser-grained sediment, deposited as the current wanes in strength, produces a structure known as a *cast*. This is a negative impression of the original structure, formed in the present case on the bottom of the overlying bed. This impression is eventually exposed as a result of differential erosion, which tends to remove the muddy sub-stratum from the base of the overlying bed, leaving it exposed to view.

Flutes vary from a few centimetres, or even less, up to nearly a metre or so in length. The larger examples are usually associated with thicker beds of coarser-grained sediment. They are typically asymmetrical in shape, with steep bulbous noses pointing up-current, while they merge down-current with the adjacent bedding-plane. Flutes are therefore excellent indicators of palaeocurrent direction, since they allow the sense of flow to be determined. For example, **41** shows that the currents flowed from the right across the bedding-plane as it is now seen.

Flute-casts are usually filled with coarser-grained sediment than the bulk of the overlying bed. Occasionally, the sedimentary infill shows evidence of cross-bedding, inclined down-current. In plan view, they vary from highly elongate forms, to triangular shapes with flaring sides, as shown in **40**. Some flute-casts mimic the shape of a horseshoe, as shown in **42**, while others have a somewhat twisted shape, giving rise to *corkscrew flute-casts*.

The erosional origin postulated for flute-casts can be confirmed wherever they are seen in cross-section to cut across the bedding of the underlying shale or mudstone. Occasionally, this is expressed by the presence of a series of small-scale steps or terraces, developed along the sides of the flute-cast itself, marking the bedding of the underlying rocks. Such a feature is produced by the differential erosion of the underlying sediment as a consequence of slight differences in lithology between the individual layers.

The occurrence of flute-casts in the form of sole structures allows them to be used as indices of stratigraphic order in turbidite sequences.

Figure 40 *Flute casts exposed on the under-surface of a greywacke bed. They represent a whole series of erosional hollows cut into an underlying sub-stratum of finer-grained sediment, which are then filled with coarser-grained sediment, forming the greywacke bed itself. Cowpeel Bridge (NT 315311), Peebleshire, Scotland. (Field of view c 1m)*

Figure 41 *Flute casts seen in a closer view of the under-surface of the greywacke bed shown in 40. They form heel-like protuberances on the bedding-plane forming the base of this greywacke bed, facing up-current. Cowpeel Bridge (NT 315311), Peebleshire, Scotland. (Field of view c 20cm)*

Figure 42 *Horseshoe flute-cast showing its typical form with a rather rounded nose, behind which a rim extends down-current to form a crescent, surrounding a central area of lower relief. Same locality as 40 and 41. Cowpeel Bridge (NT 315311), Peebleshire, Scotland. (Field of view c 20cm)*

Longitudinal Scours

As well as flute-casts, the scouring of a muddy sub-stratum by a turbidity current can result in what are known as *longitudinal scour-marks*, which are formed in most cases on rather a small scale. Since it is again a mirror image of the original structure, the ridges shown in **43** represent flat-bottomed troughs which were cut into the muddy sub-stratum by the scouring action of turbidity currents. They are commonly spaced up to a few centimetres apart, with a relief of only a few millimetres. The intervening furrows are cuspate in form, since they represent the casts of sharp-crested ridges which once separated the flat-bottomed troughs from one another on the muddy sub-stratum.

Bounce, Prod and Skip Marks

Mudstone fragments and plant and animal remains may be carried along as *tools* by the turbidity current. If they come into occasional contact with the sedimentary interface, over which the current is flowing, they produce what are known generally as *tool-marks*. Again, such structures are preserved in the form of negative impressions, occurring as casts on the base of the overlying bed.

Bounce-marks are discontinuous features, often elongate and aligned parallel to one another, which are reasonably symmetrical along their lengths. **44** shows a series of such marks, developed as sole structures on the base of a greywacke bed. *Prod-marks* differ from bounce-marks by a greater degree of asymmetry along their lengths, so that one end is blunter and more clearly defined than the other, as shown in **45**. They are formed by fragments which, approaching the sedimentary interface at a low angle, dig down into the muddy sub-stratum before they are pulled out more steeply by the force of the current, perhaps turning over as they do so.

Skip-marks occur wherever bounce-marks are so evenly spaced along a particular line that they could only have been formed by a single object, bouncing along the sea-floor in a very regular fashion. All these tool-marks show a preferred orientation, lying parallel to one another, from which the direction of flow of the palaeocurrent can be measured. However, it is often much less easy to determine the actual sense of flow, using the degree of asymmetry that is shown along their lengths. However, prod-marks typically show the greatest amount of relief where they face down-current, becoming deeper, blunter and more clearly-defined in this direction. They differ in this respect from flute-casts, so these two types of structure must be clearly distinguished from one another wherever they are used to determine palaeocurrent directions.

Figure 43 *Longitudinal scour-marks forming a markedly elongate series of ridges and furrows, all lying parallel to one another on an under-surface of a greywacke bed. They can be used to determine the direction of flow of a current, but not its sense. Hartland Quay (SS 224247), Devon, England.*

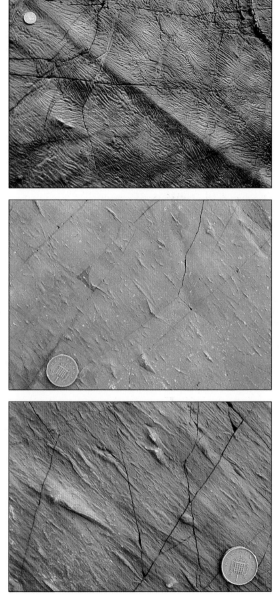

Figure 44 *Bounce-marks formed by the bouncing impact of some object on a muddy sub-stratum, which are then preserved as casts on the under-surface of an overlying bed of fine-grained greywacke. Widemouth Bay (SS 196014), Cornwall, England.*

Figure 45 *Prod-marks on the under-surface of a greywacke bed, showing how they form a series of somewhat asymmetrical impressions, which actually face down-current in the opposite direction to flute-casts. Widemouth Bay (SS 196014), Cornwall, England.*

Groove- and Chevron-casts

Objects dragged along by a current can gouge out grooves in the muddy substratum, which may then be preserved once coarser-grained sediment is deposited on top, so forming *groove-casts* on the base of the overlying bed. As impressions of the original structure, they typically occur as elongate ridges, standing proud of the surrounding surface, as shown in **46**, while these ridges may have a striated form, as seen in **47**. Occasionally, these minor striations on the sides of a groove-cast are somewhat twisted, as if the object concerned in their formation had rotated as it was dragged along by the current.

Groove-casts are very regular structures, when viewed along their lengths, showing little change in profile within the confines of a single exposure. However, although it is rarely observed in the field, groove-casts do eventually die out, sometimes in a rather abrupt manner, but usually more gradually, so that they simply merge with the surrounding bedding-plane. Very occasionally the object that actually gouged out the groove may be found at its end.

Groove-casts vary in width from a few millimetres, up to many centimetres, with the larger examples found associated with the thicker and coarser-grained beds of sandstone. They are often seen together with bounce and prod-marks, but not flute-casts. This suggests that they form once the force of the current responsible for their formation starts to wane.

Groove-casts are important indicators of palaeocurrent direction, although the sense of flow cannot be ascertained. However, although they usually occur parallel to one another, there is frequently more than one set present on a single bedding-plane. **48** shows the under-surface of a greywacke bed, on which two sets of groove-casts can be seen. Note how the cross-cutting relationships between these structures allow the two sets to be dated with respect to one another.

Chevron-casts are similar to groove-casts in that they are elongate marks preserved as sole structures. However, as the name suggests, they occur as lines of V-shaped corrugations, all pointing in the same direction. They are formed by the object concerned puckering up the muddy substratum as it is dragged along by the current. The chevron pattern formed by the corrugations must therefore point down-current, the sense of flow the palaeocurrent to be determined.

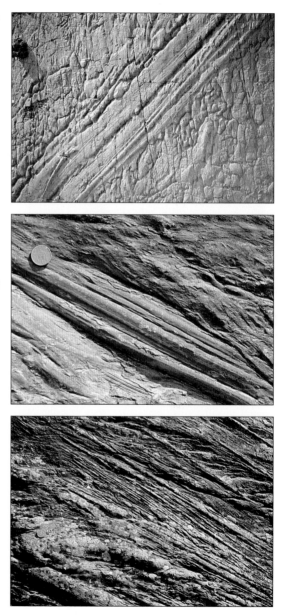

Figure 46 *Groove-cast on the under-surface of a greywacke bed, showing a rather flat profile typical of such structures, which are mostly formed by fragments of mudstone and other less common objects being dragged across a muddy substratum. Widemouth Bay (SS 198035), Cornwall, England. (Field of view c 80cm)*

Figure 47 *Groove-cast showing a series of second-order grooves and striations inscribed on a smaller scale along its length, as if the object responsible for gouging out the surface was angular with sharp corners. Cracklington Haven (SX 139966), Cornwall, England.*

Figure 48 *Under-surface of a greywacke bed showing two sets of groove-casts. Evidently, the set trending continuously across the field of view from top left to bottom right is later since it cuts across the other set, trending closer to the horizontal, and best seen in the top right-hand corner. Cracklington Haven (SX 139966), Cornwall, England.*

Post-depositional Structures

Load-casts

This post-depositional feature is found as a sole structure on the under-surfaces of sandstone beds, deposited on top of a muddy sub-stratum, so that it acts as another useful index of stratigraphic younging. Mud is laid down as a very soft sediment, capable of plastic flow, which has a relatively low density owing to the large amount of interstitial water trapped within its pore spaces. The density gradually increases during its consolidation as the pore water is expelled, so allowing compaction to take place. Sand, however loosely-packed, is deposited as a sediment of much greater density. This means that a condition of *gravitational instability* exists wherever coarser-grained sediment, such as sand or even silt, is laid down on top of mud, which remains soft and incoherent. The result is that the coarser sediment tends to sink unevenly into the underlying mud, so forming *load-casts*. They are exposed to view as sole structures, wherever erosion has stripped away the underlying shale or mudstone from the base of the bed so affected, as shown in **49**. Load-casts are the result of what has been termed *soft-sediment deformation*. Depending on the amount of differential subsidence, they can vary in shape from very slight bulges,

lacking any particular orientation, to deeply-rounded masses, which may become completely detached from the base of the overlying bed.

It is a characteristic feature of load-casts that their present form is not the result of erosion prior to the deposition of the overlying bed. This means that they should not be called casts, strictly speaking, since they do not occur as an infilling of a pre-existing feature. Even so, the relief shown by flute- and groove-casts, and other erosional structures of this sort can be accentuated by the effects of load-casting.

Where they are not keyed into earlier structures in this way, load-casts typically show a total lack of any obvious alignment, as shown in **50**.

Where load-casts occur sufficiently close together, they may isolate tongue-like masses of the underlying sediment, which penetrate upwards into the overlying horizon of coarser-grained sediment along its base to form what is known as *flame structures*. If these plumes of finer-grained sediment actually break across the bedding of the overlying horizon as they do so, they form upward injections of sedimentary material into this coarser-grained bed, as shown in **51**.

Figure 49 *Load-casts forming very irregular protuberances on the under-surface of a greywacke bed, which must have been laid down originally on a flat-lying surface of deposition. Hartland Quay (SS 224247), Devon, England. (Field of view c 1.2m)*

Figure 50 *Load-casts forming a polygonal pattern on the under-surface of a greywacke bed. They are separated from one another by tongues of darker sediment, penetrating upwards from the underlying bed, now mostly removed as the result of differential erosion. Widemouth Bay (SS 198035), Cornwall England.*

Figure 51 *Flame structure formed by a cuspate tongue of shale which has been forced upwards to penetrate an overlying horizon of silty sediment, while this sediment has sunk itself unevenly into the underlying shale to form the load-casts on either side of this sedimentary injection. Bude (SS 203073), Cornwall England.*

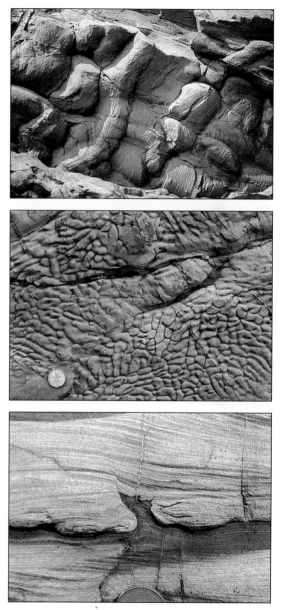

Ball-and-pillow

The load-casting of a sandstone bed can result in extreme cases in the formation of totally isolated masses of sandstone, which must have sunk into the underlying shale or mudstone while it was still soft and unconsolidated. This may occur wherever sand is deposited very rapidly over highly plastic mud, which did not have enough time for any compaction to take place as the result of dewatering. It has also been argued that earthquake shocks may have converted the overlying sediment into a quicksand, so allowing it to founder into the underlying mud. Whatever the cause, the result is known as *ball-and-pillow structure*. It commonly affects only the lower part of a sandstone bed, which is otherwise not disturbed in this way.

The sandstone masses typically occur as rounded balls or more elongate pillows of coarser-grained sediment, completely surrounded by shale or mudstone. They vary in size from a few centimetres, up to roughly a metre across. **52** shows a typical example of a sandstone pillow. Similar structures are formed if slab-like masses of sandstone have foundered into their muddy sub-stratum, forming saucer-like bodies, often with overturned rims. The internal structure of these masses allows them to be used as indices of stratigraphic younging, since it has a form which is always concave-upwards.

Although the sunken masses of sandstone are usually completely isolated within the underlying mudstone, occasionally a connection is preserved with the overlying bed. For example, **53** shows such a sandstone bed, marked by ball-and-pillow structure towards its base, which overlies a dark mudstone. Towards the centre of the field of view, a large mass of sandstone has evidently foundered into the underlying mudstone, forming a pendent in the form of a column, so that it is still linked with the overlying bed of sandstone.

54 shows that ball-and-pillow structure can also be developed within sandstone beds alone, without affecting the underlying rocks. The sandstone balls are themselves surrounded by a rather structureless rock of a similar lithology. It is then rather difficult to account for this sort of sedimentary structure solely in terms of gravitational instability without invoking another cause, such as the effects of dewatering on the sandstone, perhaps triggered by earthquake shocks.

Figure 52 *Ball-and-pillow structure in sandstone, showing a typical example of a sedimentary pillow. The internal bedding within this mass of sandstone is conformable with its outer surface, often forming a basin-like structure with a rim that is turned up-and-over on itself. Wiseman's Bridge (SN 151063), Dyfed, Wales. (Field of view c 1m)*

Figure 53 *Columnar mass of sandstone forming a pendent-like structure which, although still retaining a connection with the overlying bed of sandstone, has obviously sunk into the underlying mudstones under the influence of gravity. Broad Haven (SM 860142), Dyfed, Wales. (Height of section c 3m)*

Figure 54 *Ball-and-pillow structure in a single bed of sandstone. This example is perhaps not typical, since the sandstone balls are farther apart than is usual in such cases. Amroth (SN 156068), Dyfed, Wales.*

Convolute Bedding

Convolute bedding is a distinctive kind of internal deformation seen within many sandy or silty beds that have been laid down very rapidly. As shown in **55**, it takes the form of rather irregular contortions within the bed itself, which is otherwise not affected by the deformation. Although its origin is somewhat uncertain, it appears to affect sediment in a state of partial liquefaction caused by its very rapid deposition, so that this structure is most likely to be formed by the upwards escape of water during the very early stages of compaction. The cuspate form of the contortions allows convolute bedding to be used as an index of stratigraphic younging. The contortions die out downwards so that the base of the bed is not affected. This may also occur towards the top of the bed, although it is commonly found that a surface of erosion is developed in this position, truncating the internal contortions within the bed itself. This can also be used as an index of stratigraphic younging.

Slump Structures

Unlike the post-depositional structures already considered, which are the result of vertical movements under the influence of gravity, *slumps* are gravity-induced structures showing a large component of horizontal transport. They are formed by sedimentary masses sliding down what is known as a *depositional slope*. This need be only a few degrees, or even less, for slumping to affect soft sediment containing a large amount of pore water.

Slumps vary considerably in size and structural complexity. However, they typically rest on a basal plane, along which the slumped mass has detached itself more or less from the underlying rocks, while they are generally overlain by bedded sediments, which were subsequently deposited on top of the slumped mass. This results in a series of *slump-sheets*, separated by normal sediments, which form a part of the stratigraphic sequence. These sheets vary in thickness from just a few metres, or occasionally less, upwards to many tens of metres. The thicker the sheet, the wider an area that it covers.

56 shows the typical appearance of a slump-sheet. The bedding dips steeply to the left, and youngs in the same direction. The oldest rocks occur towards the right, where they form a normally bedded sequence. The slump itself forms a jumbled-up mass of disrupted and distorted beds, lacking anything in the way of a regular structure. Other slump-sheets consist of bedded sediments which have folded as a result of the movements, often accompanied by a certain amount of faulting.

57 shows such a *slump-fold*, perhaps best termed a slump-roll in the present circumstances, since most examples show a much greater degree of deformation and down-slope movement than this illustration suggests.

Figure **55** *Convolute bedding in a sandstone, showing how the internal contortions form a series of rather narrow ridges with sharp crests, separated from one another by much broader trough-like features, across which the bedding can be traced without a break. Corrie (NS 025428), Isle of Arran, Scotland.*

Figure **56** *Slump sheet formed by the break-up of several beds of fine-grained sandstone as they moved as a coherent mass down a slight slope under the influence of gravity. The stratigraphic section youngs towards the left. Baggy Point (SS 426401), Devon, England.*

Figure **57** *Slump fold affecting not just a thin bed of red sandstone, but also the underlying rocks, which was formed in response to only a slight amount of movement down a depositional slope, inclined in this instance towards the left. Clachtoll (NC 037272), Sutherland, Scotland.*

Sand Volcanoes

Although they occur only rarely in the geological record, *sand volcanoes* are found on bedding-planes wherever the upward injection of quicksand has resulted in its eruption at the surface. **58** shows a typical example, conical in shape with a central depression. They range in diameter from several centimetres up to a few metres across, forming a feature of relatively low relief. They generally occur immediately above highly disturbed beds showing the effects of slumping, loading and convolute bedding, suggesting that they are the result of liquefaction of water-saturated sediment.

Sandstone Dykes and Sills

Occasionally, sheet-like bodies of sandstone are seen to cut across the bedding of sedimentary rocks. They are formed by the injection of quicksand in a highly mobile condition into fissures, which must have opened soon after the surrounding rocks were deposited. The terminology applied to the structures formed as a result follows that used to describe igneous intrusions. **59** shows that *sandstone dykes* are sheet-like bodies with parallel sides that cut across the bedding of the surrounding rocks at a high angle. Two dykes can be seen in the present instance, derived from the underlying bed of sandstone. This is commonly found to be the case, so that the liquefaction of the sandstone now occupying the dyke was probably the result of an earthquake shock or some other disturbance, which converted the underlying sandstone into a quicksand. Sandstone dykes are rarely more than a metre in width, often much less, while they usually cannot be traced for any great distance. They often show the effects of differential compaction, since they are usually surrounded by finer-grained sediment which can undergo a greater degree of compaction under the weight of the overlying rocks. This can result in sandstone dykes showing a buckled form (see **59**). Another example of a sandstone dyke is shown in **60**, showing much the same features. Note that all these dyke-like structures differ from Neptunian dykes (see **81**), which are filled by sediment from the surface.

Sandstone sills are sheet-like bodies which have been injected along the bedding, so that they are difficult to distinguish from ordinary beds of sandstone. However, cross-cutting relationships may be present locally, while they generally lack the internal bedding and other depositional structures typical of normal sandstones. If they can be recognized, sandstone sills present clear evidence for the injection of quicksand in a highly mobile condition.

Figure 58 *Sand volcano formed by the penecontemporaneous eruption of quicksand at the surface. Some examples show lobate masses of sediment which have evidently slumped down the sides of the cone in a state of liquefaction. Goleen Bay, County Clare, Ireland. (Photograph by G.M. Harwood)*

Figure 59 *Sandstone dykes cutting vertically through a shaly horizon lying between two greywacke beds, and derived from the underlying bed. Note how both dykes are buckled to a slight extent in response to the vertical compaction that has affected the surrounding shales to a greater degree. New Quay (SN 385604), Dyfed, Wales.*

Figure 60 *Sandstone dyke cutting the same horizon as the previous photograph, and showing much the same features. New Quay (SN 385604), Dyfed, Wales.*

Shrinkage Cracks

Desiccation cracks, often called sun-cracks, are a common and well-known feature of sedimentary rocks, formed by the contraction which occurs when a muddy sediment dries out on exposure to the air. This produces a tension, acting uniformly in all directions within a horizontal plane, and diminishing downwards in its intensity, away from this surface. A polygonal network of vertical cracks, tapering downwards from the surface, is generated as a response to this horizontal tension. The polygons are often but not always six-sided, while they vary from several centimetres up to a metre or so across.

Desiccation cracks are preserved in the geological record wherever coarser-grained sediment, usually sandstone, fills in the vertical cracks. They are therefore seen as casts on the base of the overlying bed, where they form a useful index of stratigraphic order. However, the overlying bed is commonly enough stripped off as the result of erosion, so exposing the underlying bedding-plane, as shown in **61**.

Shrinkage cracks can also form underwater, when they are known as *synaeresis cracks* (Latin: *synaeresis*, contraction). Their mode of formation is not clearly understood, but it appears to involve the shrinkage of muddy sediment in response to changes in salinity. **62** shows how they are usually preserved on bedding-planes, in exactly the same way as desiccation cracks. However, they do not form polygonal networks. Instead, they tend to form rather irregular patterns, while they are usually developed on a much smaller scale. The individual cracks may, however, be roughly parallel to one another, or they may form radiating networks, cutting across one another as a consequence. They also tend to be lenticular in plan, wedging out along their lengths rather than joining up with one another.

63 shows a whole series of synaeresis cracks in vertical cross-section, showing the effects of *compaction folding*. The bedding is formed in this case by thin layers of light-coloured siltstone, separated from one another by calcareous mudstone, orangey-brown in colour. Synaeresis cracks, filled from above with silty sediment, penetrate downwards into the mudstone. A clear example is seen towards the centre of the field of view. Compaction folding occurs because the silt-filled cracks were originally flanked by muddy sediment which could undergo much more compaction than the silty material itself. The synaeresis cracks then became folded in order to accommodate the vertical reduction in thickness of the sedimentary beds as a whole, since the silty material filling these cracks could not undergo as much compaction as the surrounding sediment.

Figure 61 *Desiccation cracks filled with relatively coarse-grained sediment, and separated from one another by calcareous mudstone, forming a polygonal network on the upper surface of a bedding-plane. Clairdon Head (ND 138700), Caithness, Scotland.*

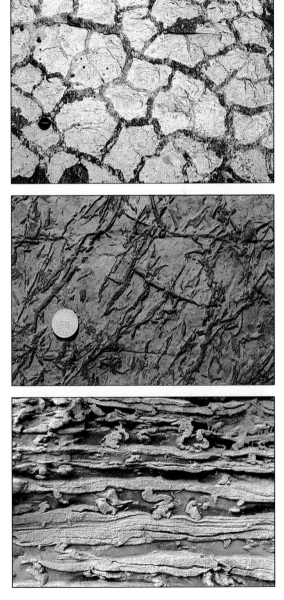

Figure 62 *Synaeresis cracks filled with silty sediment, exposed on a surface of calcareous mudstone. Note the marked difference between the somewhat irregular pattern formed by these cracks, and the polygonal networks typical of desiccation cracks, as shown in the previous photograph. South Head (ND 376497), Caithness, Scotland.*

Figure 63 *Synaeresis cracks as seen in vertical cross-section, each penetrating downwards into a calcareous mudstone of orangey-brown colour from an overlying layer of light-coloured siltstone, showing the effects of compaction folding. South Head (ND 376497), Caithness, Scotland.*

Tracks, Trails and Burrows

Sediment may also be disturbed by the activity of living organisms, prior to its final consolidation. The record of this activity is preserved in the form of tracks, trails and burrows. Such structures are known collectively as *trace fossils* to distinguish them from *body fossils*, which are the actual remains of living organisms, even if only preserved as casts. Although the forms shown by trace fossils can often be interpreted in terms of a particular mode of life, it is often difficult to identify the fossil species involved in their formation. Trace fossils are therefore classified according to their morphology.

Trace fossils can be divided into two broad categories according to whether they are found as *tracks* and *trails* on the surfaces of bedding-planes, or whether they occur as *burrows* excavated into the underlying sediment. Apart from footprints and other discontinuous marks, crawling traces tend to occur as straight or sinuous trails, such as shown in **64**, while grazing traces usually have a more complex pattern which results from the organism feeding in a systematic manner. All these traces are usually imprinted on a muddy sub-stratum. They are most commonly preserved as casts on the base of a coarser-grained bed, which was subsequently deposited on its top. Such structures provide a useful index of stratigraphic order.

Burrows can vary widely in shape from relatively simple forms to highly complex structures, lying at a variety of different angles to the bedding. In fact, some burrows occur along particular horizons within the sediment, so that they are eventually exposed on bedding-planes. They are then difficult to distinguish from surface traces without specialized knowledge. As shown in **65**, burrows are usually recognized in vertical cross-section at right angles to the bedding. They can then be seen by the contrast in lithology which exists between their sedimentary infilling and the surrounding rocks. This is commonly emphasized by the fact that burrows are often filled with rather structureless material. However, some organisms produce an internal lamination within this material as the result of backfilling after their passage through the burrow itself. One effect of burrowing organisms is that the surrounding sediment becomes increasingly disturbed. If sufficiently intense, this may eventually destroy any bedding within the sediment, as the result of what is then known as *bioturbation*.

On occasion, burrows can be used as evidence of stratigraphic younging. **66** shows a trace fossil known as *Monocraterion*. The pipes are filled with somewhat structureless sediment, while the bedding in between is deflected downwards against their margins. The pipes also have a conical shape, opening upwards, as indicated by their common name of trumpet pipes, so that the direction of stratigraphic younging can be recognized.

Figure 64 *Trace fossil forming a series of simple trails without any obvious ornament, made by several invertebrate organisms crawling across the upper surface of a sedimentary bed soon after its deposition. St Monance (NO 534018), Fife, Scotland.*

Figure 65 *Organic burrows as seen in vertical cross-section, filled with silty sediment, and showing a marked contrast in lithology with the surrounding material. Note how these burrows cut across the fine laminations defining the bedding of this darker and finer-grained sediment. St Monance (534018), Fife, Scotland.*

Figure 66 *Trumpet pipes (Monocraterion) as seen in vertical cross-section in a fine-grained quartzite, showing how the form of these upward-opening burrows can be used as an index of stratigraphic younging. Beinn Heilam (NC 466622), Sutherland, Scotland.*

Diagenetic Structures

Nodules and Concretions

The segregation of chemical material in a sediment soon after its deposition leads to the formation of *nodules* (Latin: *nodus*, a knot) and *concretions* (Latin: *concretus*, grown together). If a distinction is made, nodules occur as irregular masses, while concretions are spherical or ellipsoidal structures, often flattened in the plane of the bedding. They are mostly formed by fine-grained carbonate minerals such as calcite, dolomite, ankerite and siderite, particularly in shales and mudstones, while silica is another common mineral occurring chiefly in the form of flint and chert nodules in limestones, as shown in **67**. However, a variety of other minerals can be found, including pyrites, gypsum, anhydrite, barytes, haematite and limonite.

The segregations formed by these various minerals then form discrete bodies, differing in lithology from the surrounding rock. They vary in size from only a few millimetres up to a metre in diameter, or even more. They show a great variety of shapes, although it is commonly found that they occur at particular horizons within the stratigraphy. Although some concretions are nucleated around fossils, while others may be associated with burrows, most nodules and concretions are not related to any pre-existing structure in the rock.

Nodules and concretions can form at various stages during the course of diagenesis, as the sediment undergoes compaction and consolidation. Some nodules are formed late in the diagenetic history, when the sediment cannot undergo any further compaction. They may be recognized as late diagenetic nodules by the way in which the bedding can be traced through each nodule without any deflection. Other concretions are formed early in the diagenetic history, before much compaction has been achieved. **68** shows how these early diagenetic concretions can be recognized by the way that the bedding is deflected as it passes around the concretion.

Some concretions form what are known as *septarian nodules* (Latin: *septum*, a barrier), in which a complex array of radial and concentric cracks are filled with crystalline material, often in the form of calcite or siderite. These cracks apparently form as a result of the contraction which accompanies the dewatering of the finer-grained material that otherwise forms the concretion. Other concretions occur in limestones as *geodes*, which are seen in the form of cavities, surrounded by a layer of extremely fine-grained silica, known as chalcedony, and often lined internally with extremely well-developed crystals of quartz and calcite.

69 shows a structure known as *cone-in-cone*. It is found as a thin layer around some concretions, while it also occurs along the bedding of some shales. The present example marks the contact between shale and an overlying bed of greywacke. It is formed by fibrous calcite, less commonly siderite, which occurs as a series of closely-packed cones, sometimes separated from one another by clay-films.

Figure **67** *Chert nodules forming a series of rather irregular masses of siliceous material, embedded in a much darker limestone. Durness (NC 370626), Sutherland, Scotland.*

Figure **68** *Pre-compaction concretions, formed in silty sediment at an early stage of diagenesis. Note how subsequent compaction has caused the bedding in the surrounding rocks to be deflected around these concretions. Wiseman's Bridge (SN 154065), Dyfed, Wales.*

Figure **69** *Cone-in-cone structure affecting a concretion which has formed along a contact between shales and an overlying bed of greywacke. It is a secondary structure probably formed in response to recrystallization under the weight of the overlying rocks. New Quay (SN 385604), Dyfed, Wales.*

PART II
IGNEOUS AND METAMORPHIC ROCKS

Pillow lava formed by the submarine eruption of molten lava. Newborough Warren, Wales.

NATURE OF IGNEOUS ACTIVITY

Although the Latin root (*ignis*, a fire) implies the act of combustion, *igneous rocks* are more the product of melting, deep within the earth's crust or upper mantle. The molten material formed as a result is known as *magma*. It is essentially a melt of silicate minerals at a high temperature, consisting of silica SiO_2, togther with lesser amounts of alumina Al_2O_3, the iron oxides FeO and Fe_2O_3, lime CaO, magnesia MgO, and the alkalis, soda Na_2O and potash K_2O, as its main constituents. Dissolved under pressure are volatile substances, chiefly water H_2O, but including some other gases. The temperature of such a magma, rich in volatiles, varies between 600°C and 1200°C depending on its composition, so that it forms an incandescent mass.

Once formed at depth, magma displays a marked tendency to rise towards the surface, breaking through the pre-existing rocks of the earth's crust as it does so. The most obvious manifestation of this activity occurs wherever the magma actually reaches the surface, where it is erupted as molten lava from a volcano. Each volcanic eruption produces a *lava-flow*, which travels a certain distance from its source before it loses so much heat that it starts to solidify into an igneous rock. Lava-flows therefore form one category of the igneous rocks, occurring in the form of *igneous extrusions*. However, the magma in ascending towards the earth's surface may cool down so much against its surroundings that it starts to solidify while still within the earth's crust. This produces another category of igneous rocks, which are known as *igneous intrusions*. They occupy erstwhile spaces within the earth's crust, which were originally the pathways for the magma in its ascent towards the surface. Such rocks occur in contact with the pre-existing rocks of the earth's crust, which are known in general as the *country-rocks* to the igneous intrusion.

A broad distinction can therefore be made between extrusive lava-flows, erupted at the earth's surface, and igneous intrusions, which penetrate the pre-existing rocks of the earth's crust. Although there are marked differences in lithology between extrusive and intrusive rocks, they do show certain features in common with one another. In particular, all igneous rocks tend to be rather massive, showing a general lack of structural features apart from jointing. However, the volatile substances dissolved in magma tend to be released as the pressure drops once the magma reaches shallower levels, close to the earth's surface. This results in the formation of gas-bubbles or *vesicles* (Latin: *vesica*, a bladder), which are particularly common in lava-flows. They are often filled with secondary minerals, forming what are then known as *amygdales* (Greek: *amygdule*, an almond). The vesicles within a fluid lava-flow may be aligned in sheets parallel to the direction of flow, while the larger vesicles in the upper parts of the flow are often drawn-out and stretched in response to its continued movement.

Igneous Textures

A broad distinction is made first between *glassy rocks* which are rapidly quenched melts lacking any crystalline structure, and *crystalline rocks* in which the various silicate minerals present have crystallized out from the melt as it cooled. *Crystallization* involves the nucleation and subsequent growth of particular minerals, which eventually form the rock as a whole. The onset of crystallization takes place once the temperature of the magma falls below its freezing point. The textures developed as a result depend to a very considerable extent on the rate of cooling.

If the magma cools rather slowly, nucleation occurs to only a limited extent. This allows just a few crystalline "seeds" to form, which eventually grow into relatively large crystals, so producing a coarse-grained rock. The reverse happens if cooling is rapid. Many crystalline "seeds" are then produced, but these can only grow to a limited extent before crystallization as a whole is complete. This results in a fine-grained rock which may indeed retain a glassy groundmass to whatever crystals are present, if cooling was particularly rapid. A *porphyritic texture* is the result if there are a few relatively large crystals, known as *phenocrysts*, set in a much finer-grained or even glassy matrix. Any platy or elongate phenocrysts present within an igneous rock may show a *preferred orientation* with their longer dimensions parallel to the direction of flow shown by the magma. Glassy rocks may subsequently undergo devitrification, after they have solidified as a supercooled liquid from a melt, leading to the subsequent growth of fine-grained crystals in the solid rock.

The rate of cooling of a magma depends on how quickly heat can be lost to its surroundings. This depends in its turn on the surface area of the igneous body, taken in comparison with its volume. Although there are many exceptions to the rule, this means that fine-grained and glassy rocks usually occur as lava-flows and *minor intrusions*, while coarse-grained igneous rocks are typically found on a much larger scale as *major intrusions*.

Chemical and Mineralogical Composition

Apart from rare exceptions, igneous rocks mostly consist of more than one mineral species. Such *rock-forming minerals* are nearly all silicates, given the high proportion of silica in the earth's crust. Apart from quartz itself, the rock-forming silicate minerals are chemical compounds of silica occurring in combination with the metallic oxides commonly found in the igneous rocks. The silica was originally regarded as playing the role of an acid, while the other metallic oxides were thought to act as bases. Igneous rocks, such as *granite*, which are rich in silica, soda and potash, but poor in iron oxides, magnesia and lime, have traditionally been known as *acid rocks*. They consist of light-coloured minerals such as quartz and the alkali feldspars, along with minor amounts of dark-coloured minerals such as biotite and hornblende. They can be compared with what are known as the *basic rocks*, such as *basalt*, which are relatively poor in silica and the alkalis, but rich in iron oxides, magnesia and lime. Quartz is often lacking in such rocks, which consist of lime-rich feldspars together with dark-coloured minerals such as pyroxene and olivine, in roughly equal proportions. The *intermediate rocks* are halfway in composition between acid and basic rocks, while rocks very poor in silica are known as *ultrabasic*. The latter consist almost entirely of dark-coloured minerals such as pyroxene and olivine. The further classification of igneous rocks is a subject well beyond the scope of this book. However, igneous rock-names are often used rather informally in the field, and this is a practice that we shall follow in this book, using the terminology just introduced.

Pyroclastic Rocks

Volcanoes can also produce fragmental matarial as a result of explosive activity, forming what are known as the *pyroclastic* rocks. This type of volcanic activity is commonly associated with the more acid magmas, which have so high a viscosity that the volcanic gases cannot escape at all easily from the magma as it moves closer to the surface, where the pressure is obviously less. Although molten lava can be ejected as volcanic "bombs", much fragmental material is thrown out as solid rock, torn from the walls of the volcanic vent by the force of this explosive activity. Although some falls back into the vent, or it may never have reached the surface, much is distributed over wide areas around the volcano. Clearly, the largest fragments will only travel a short distance before they fall to the ground, while the finest particles will be carried away for much greater distances before they finally settle out of the atmosphere. A pyroclastic rock is known as an *agglomerate* if it has a particle size exceeding 64mm, at least according to some definitions. If the rock is finer-grained than an agglomerate, it is called an *ash* or a *tuff*. Describing a pyroclastic rock as a volcanic ash only implies that its fine-grained nature is the result of extreme fragmentation, not that the rock is itself a residue of combustion. The particles in a volcanic ash can include glass shards and crystal fragments, derived from the magma itself during the course of its crystallization.

Ash-falls and Ash-flows

Pyroclastic rocks formed by the explosive ejection of fragmental material from a volcanic vent are known as *ash-falls*. Commonly, they show bedding or stratification, produced by fluctuations in the force of the volcanic explosions, perhaps augmented as far as the finer-grained deposits are concerned by changes in atmospheric conditions. They may well accumulate under water, where they may be affected by reworking to form what is essentially a sedimentary rock. This mode of eruption is very different from that known as an *ash-flow*. This occurs as a turbulent suspension of rock fragments, crystal particles, pumice fragments and glass shards, all carried along by very hot gas. An ash-flow is apparently produced by the explosive vesiculation of acid magma, as suggested by the presence of pumice fragments and glass shards in the resulting deposit, which is known as an *ignimbrite*. It travels close to the ground as a coherent but highly mobile mass, eventually coming to rest and filling in any hollows in the topography. The pumice fragments and glass shards in this deposit, while still very hot, not only collapse under the weight of the overlying material, but may also become welded together, so forming what is known as a *welded tuff*.

METAMORPHIC PROCESSES

Metamorphic rocks are formed from pre-existing rocks wherever they have been so altered by various processes taking place in the solid that they have effectively been changed into rocks of a different lithology. Apart from *dislocation metamorphism*, which results in the mechanical breakdown of rocks affected by faulting, most metamorphic processes involve the recrystallization of existing minerals in the rock, together with the nucleation and subsequent growth of new minerals. Such *metamorphic reactions* occur in response to changes in temperature and confining pressure, which affect the pre-existing rocks at some depth with the earth's crust. Although extreme metamorphism may eventually result in partial or even complete melting in some cases, metamorphic reactions generally take place in the solid rock. However, they may be influenced by the presence of pore-fluids, particularly water. These reactions typically result in the formation of what are known as *metamorphic minerals*, such as garnet, cordierite, andalusite, kyanite and sillimanite. These minerals commonly occur in the form of *porphyroblasts* (Greek: *blastos*, a bud), forming relatively large crystals set in a much finer-grained matrix.

Contact metamorphism occurs as the result of thermal alteration, which often affects the country-rocks in the immediate vicinity of an igneous intrusion. The degree of alteration depends on the temperature of the magma, the size of the igneous intrusion, and the total amount of magma that has flowed past its contacts. Since the heating effect of an intrusion falls off as the country-rocks are traced away from its contact, the altered rocks usually form what is known as a *metamorphic aureole* around the intrusion, varying in width according to the circumstances.

Regional metamorphism differs from contact metamorphism in that it typically affects wide zones of the earth's crust, known as mobile belts, which correspond to the underground roots of folded mountain-chains, such as the Alps or the Himalayas. Rocks undergoing regional metamorphism are often affected by relatively high temperatures, along with a confining pressure exerted by the weight of the overlying rocks. However, these zones are also the focus for intense earth-movements, causing the complex patterns of deformation typical of regionally metamorphosed rocks. This means that the metamorphic processes of recrystallization and new mineral growth take place while the rocks are also being folded and otherwise deformed, commonly as a result of repeated episodes of earth-movements.

Quartzites and Marbles

The nomenclature of metamorphic rocks is relatively simple, since it is based essentially on the textural and structural features, developed in rocks of a particular lithology. The contact metamorphism of sandstones and limestones simply results in the recrystallization of quartz and calcite, producing respectively *quartzites* and *marbles*. These are frequently coarser in grain than the rocks from which they were formed. Similar rocks are formed as a result of regional metamorphism.

Hornfels

Shales affected by contact metamorphism are generally changed into a *hornfels* if the alteration is sufficiently intense. This is a hard, compact, tough and fine-grained rock, often with a purplish tinge and a somewhat flinty appearance. Any pre-existing features such as bedding tend to be obscured or even obliterated. Cordierite and andalusite may be present as porphyroblasts, often with rather indistinct outlines, giving the rock a pitted or otherwise spotted appearance on weathered surfaces.

Slates

Shales affected by varying degrees of regional metamorphism produce entirely different rock-types in comparison with contact metamorphism. If the temperature and confining pressure remain low, the effects of deformation become paramount. The resulting rock is known as a *slate*. It can be split into very thin sheets, along planes independent of the bedding, which may then be used for roofing. This property of easy splitting arises from the presence of a *metamorphic fabric* known as a *slaty cleavage*. It is formed by the parallel orientation of the constituent minerals, particularly chlorite, sericite and muscovite. These belong to a complex group of platy minerals, which includes the micas. They all share an atomic structure which allows them to split easily along a single plane of crystallographic cleavage. The property of slaty cleavage then arises from the fact that all the micaceous minerals in a slate, which normally occur in the form of extremely fine-grained flakes, are aligned more or less parallel to one another, so allowing the rock to split most easily in this direction.

Phyllites and Schists

Slates are very fine-grained rocks in which the constituent minerals are not visible even with a hand-lens. They pass with increasing metamorphism into rocks known as phyllites and schists. *Schists* are distinctly crystalline rocks in which the individual minerals, particularly the micas such as biotite, are visible. Structurally, a *schistosity* (Greek: *schizo*, split) is developed through the parallel alignment of platy minerals, generally mica. Although this allows the rock to split like a slate, a schistosity is generally much coarser and less evenly formed than a slaty cleavage. *Phyllites* (Greek: *phyllon*, a leaf) are schistose rocks with a lustrous appearance, lacking in a slate, which do not show the degree of crystallization typical of schists. They may be distinguished from *phyllonites* which are the product of retrogressive metamorphism, whereby mineral assemblages formed at higher temperatures and confining pressures are broken down into rocks of a lower grade, looking much like a phyllite.

Gneisses

A distinction can often be made in the field between schists and gneisses. A *gneiss* is defined generally as any coarse-grained metamorphic rock, with a grain-size exceeding 2mm. It is formed in response to conditions of particularly high temperature and confining pressure, as appropriate to quite considerable depths within the earth's crust. The schistose textures generally found at lower grades of regional metamorphism are often replaced by much more even-grained textures within gneisses. Feldspar also typically makes an appearance in these rocks. It is commonly associated with quartz in the form of distinct layers, separated from one another by the other minerals in the rock. This structure produces the gneissic banding, sometimes known as a *foliation* (Latin: *folium*, a leaf) if it occurs on a particularly fine scale, which is characteristic of many gneisses.

(Above) Igneous dyke cutting metamorphic country-rocks on Shetland, Scotland. The margins of this dyke are much darker than its interior. (Below) Veins of pink granite intruded discordantly across the bedding of metamorphic country-rocks. Loch Cluanie, Scotland.

Lava-flows and Pyroclastic Rocks

Age-relationships of Volcanic Rocks

The form of any lava-flow largely depends on the viscosity of the magma as it is erupted. Acid magmas tend to be highly viscous, while basic magmas are much more fluid. This means that acid lavas form thicker flows, much more limited in their extent than basic lavas. The latter, along with lava-flows of intermediate composition, are usually thinner and more widespread. This applies particularly to basic lava-flows, which individually are rarely more than 10m in thickness, even though they may cover many square kilometres in area.

Since lava-flows and pyroclastic rocks are erupted one after another at the earth's surface, they form what are effectively stratigraphic sequences of volcanic rocks, interbedded with one another. Since the oldest rocks occur at the base of such a sequence, overlain by progressively younger deposits, the Principle of Superposition can be applied. Volcanic rocks can also be dated stratigraphically. Although fossils are very unlikely to be preserved in volcanic rocks, apart from ash-falls, they may well be found in any sedimentary rocks interbedded with lava-flows or other pyroclastic rocks. It is usually possible at least to bracket the stratigraphic age of a volcanic sequence, since it must be younger than any underlying rocks but older than any rocks deposited on top, which could well be dated by palaeontological methods.

The stratified nature of a volcanic sequence produced by the eruption of lava-flows at the earth's surface is shown in **70**. The lower ground is occupied by metamorphic rocks of an underlying basement, on top of which the lava-flows erupted. These rocks form the rather subdued topography in the foreground, beyond which the volcanic rocks are exposed in the cliffs. The volcanic sequence dips at a low angle towards the left. This can be observed because lower parts of each lava-flow, particularly its central portion, tend to resist the effects of weathering and erosion more effectively than its top. Such effects of differential weathering and erosion on a volcanic sequence are known as *trap featuring* (Swedish: *trappar*, steps).

71 shows the bedded nature of pyroclastic rocks, resulting from the presence of agglomeratic layers in a fine-grained tuff. Bedded tuffs and ashes may occur on the flanks of a volcano, away from its vent, or they may be preserved within the vent itself. Pyroclastic rocks, particularly agglomerate, tend to show initial dips, corresponding to the form of the volcanic edifice, decreasing away from their source, while such rocks may become increasingly disturbed as the result of further activity, wherever they occur within the vent itself.

The explosive ejection of large blocks from a volcano may produce a useful index of stratigraphic order, as shown in **72**. Note how the bedding is deflected underneath the block as a result of its impact on the surface.

Figure 70 *Lava-flows of Devonian age in the Glencoe cauldron-subsidence of Scotland, forming a well-stratified sequence of volcanic rocks, which were erupted on top of an old land-surface of metamorphic rocks, which now forms the rather subdued topography in the foreground. Aonach Dubh (NN 139569), Argyll, Scotland. (Height of section c 800 m)*

Figure 71 *Bedded agglomerate in a volcanic vent, showing how the bedding is defined in rather a crude fashion by the larger fragments, chiefly of lava, set in a much finer-grained tuff. Elie Ness (NT 499994), Fife, Scotland.*

Figure 72 *Bedded tuff showing how its bedding has been deflected underneath a much larger block of lava, blown into the air by the force of a volcanic explosion, to form an impact structure that can be used as an index of stratigraphic younging. Maure Vieille, Esterel Massif, Provence, France. (Field of view c 2 m)*

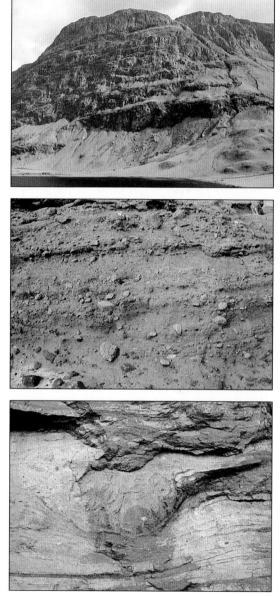

Lower Contacts of Lava-flows

Although lava-flows often rest on top of one another, their eruption may be interrupted by the deposition of sedimentary rocks, or by the formation of pyroclastic rocks. Individual flows may therefore be found interbedded not only with other lavas but also with sedimentary or pyroclastic rocks.

In flowing across a pre-existing surface, the base of a lava-flow will obviously take on the form of this surface. For example, **73** shows the base of a lava-flow which occupies a small valley cut into the underlying rocks. Note the presence of a crude columnar jointing (see **97**) arranged roughly at right angles to the cooling surface formed by the base of the lava-flow. Normally, however, it is found that most lava-flows rest as flat-lying layers in apparent conformity with the underlying sequence. Exceptions to this general rule only occur where the volcanic activity has produced some degree of topographic relief, often through the eruption of rather acid lava. Such lava, being less fluid, tends to flow for shorter distances, often ending rather abruptly with a steep face to the flow. This produces a volcanic stratigraphy which is much less regular than that found with more basic lavas.

The basal parts of relatively fluid lava-flows are often finer-grained and more vesicular than their central portions. Basic lava-flows somewhat less fluid than normal may incoporate a thin rubbly zone at their base, consisting of lava-blocks which fell off the front of the flow as it advanced. Columnar jointing may well be present above such a basal layer, commonly forming two tiers of differently-sized joints throughout the central parts of the flow itself. As the name suggests, these joints are usually vertical, although occasionally they are tilted in the direction of flow.

The heat given off from a lava-flow can also "bake" the underlying rocks if they are particularly susceptible to the effects of thermal alteration. Likewise, any vegetation overriden by the lava-flow will be charred. Very occasionally, fluid lava-flows are erupted on top of muddy sediment. **74** shows how load-casts can then be formed along the base of the lava-flow.

Pipe Amygdales

75 shows a particular form of vesicular structure which is sometimes found at the base of relatively fluid lava-flows. Note how the vesicles form pipes approximately at right angles to the base of the flow. They were produced by gas-bubbles rising vertically upwards through the lava. Their present inclination away from the vertical is a consequence of the lava flowing towards the left after the pipes were formed. Such pipes can be used as an index of stratigraphic younging whenever two or more pipes join together as they are traced upwards. This can only happen if adjacent gas-bubbles coalesce with one another as they rise through the lava.

Figure 73 *Base of a basaltic lava-flow, showing a marked discordance with the structure of the underlying rocks, owing to its eruption across a pre-existing land-surface. Bitlis Gorge, Eastern Turkey.*

Figure 74 *Base of a basaltic lava-flow, showing how load-casts can be produced as a result of differential subsidence through the eruption of relatively fluid lava on top of soft mud. Note how the vesicles close to the surface of the lava become deformed as a result of its flow. Bitlis Gorge, Eastern Turkey.*

Figure 75 *Pipe vesicles lying just above the base of a basaltic lava-flow, showing how they can be used as an index of stratigraphic younging wherever they coalesce upwards with one another to form an inverted Y-shape. Homs, Syria.*

Flow-structures and Autobrecciation

Various structures are formed by the flow of molten lava. The surfaces of very fluid flows often congeal to form a smooth, glassy skin, often somewhat vesicular, which then becomes wrinkled and billowed as the lava underneath continues to flow. This type of ropy lava is known in Hawaii as *pahoehoe*. It is rarely preserved in ancient lava-flows, presumably because it forms only a thin crust to the lava-flow. Such a fluid lava-flow in solidifying may be left with a still molten core, through which the remaining lava flows away, leaving what is known as a *lava tunnel*. An example is shown in **76**, filled with earthy material derived from weathering and erosion of the lava-flow.

More viscous lava frequently forms a surface crust, thicker and very much rougher than pahoehoe. This breaks up into separate blocks as the lava flows along underneath, so forming blocky lava. The process is often accompanied by the escape of gas from the half-congealed lava as it breaks up into fragments. Such vesiculation occurs preferentially along incipient fractures, wherever there is a slight reduction in pressure. This leads eventually to the complete fragmentation of highly vesicular lava forming what is known as *aa* lava in Hawaii. It often occurs as the very slaggy or *scoriaceous* tops to basic lava-flows, which are commonly enough preserved in the geological record.

77 shows the type of blocky structure known as *autobrecciation*, which is typically found in lava-flows which are somewhat more acid in composition. The rather angular blocks of light-coloured rock were formed by the fragmentation of an already solidified portion of the lava-flow. The darker rock surrounding these blocks represents the once still-molten lava from the same flow which eventually cemented these blocks together as the lava-flow came to rest and solidified.

Rather different structures are formed in more acid lava as a result of its flow. As shown in **78**, these structures often occur in the form of a platy *flow-banding*, which is often intensely folded in a very complex manner. Any slight variations in the nature of such a viscous lava tend to be accentuated during the course of its flow, particularly once it starts to congeal and crystallize. The differences in lithology which are developed as a result are marked by slight changes in crystallinity and mineral composition between the layers. They are best seen on slightly weathered surfaces where they are often etched out as bands differing in colour and texture. It is quite common for these lavas to undergo autobrecciation at a later stage of their emplacement, so that the broken-up blocks then show an internal flow-banding.

Figure 76 *Lava tunnel formed in the central parts of a lava-flow, through which the still-molten lava flowed away during the final stages of its consolidation. Bitlis Gorge, Eastern Turkey. (Height of section c 4m)*

Figure 77 *Autobrecciation affecting the upper part of an andesitic lava-flow, formed wherever still-molten lava has cemented together broken-up fragments of the same flow, which had already become solidified during the course of its eruption. Coire nan Beith (NN 139553), Glencoe, Scotland.*

Figure 78 *Flow banding in an acid lava-flow, formed by the streaking out of layers showing slight differences in composition and texture prior to the final consolidation of such a viscous lava. Notre Dame de Jerusalem, Esterel Massif, Provence, France.*

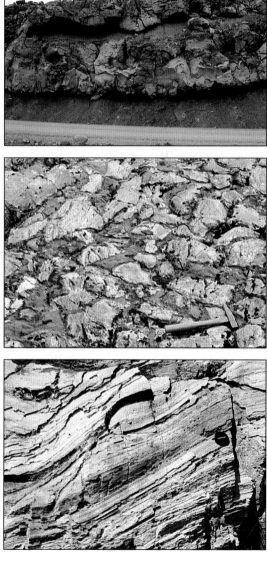

Upper Contacts of Lava-flows

Contrast with Lower Levels The uppermost parts of many lava-flows, particularly if they are more basic than acid in their composition, often present a marked contrast with their lower levels. This not only allows each lava-flow to be identified as a distinct unit, but it also provides a useful index of stratigraphic younging for the volcanic sequence as a whole. Even if a particularly scoriaceous top does not form the upper part of a lava-flow which is otherwise rather massive, even-grained and well-jointed, it is commonly found that any vesicles present in the lava-flow become larger and much more abundant towards its upper surface, as shown in **79**. Although there may also be a thin vesicular layer at the base of any lava-flow, it is rarely so well developed. The individual gas-bubbles are often filled with secondary minerals to form amygdales.

Weathered Bole The eruption of many lava-flows often occurs under subaerial conditions, forming a series of very short-lived events each separated by long periods of inactivity. If the climate is favourable, chemical weathering can then start to act upon the top of the lava-flow which has just been erupted. The products of chemical decomposition formed as a result would eventually be buried by the eruption of the next lava-flow in the volcanic sequence. Such profiles of chemical weathering found at the tops of basic lava-flows are known as *boles* wherever they take on a red coloration due to the presence of iron oxides in the decomposed rock, as shown in **80**. Note how the red coloration becomes less intense as it is traced downwards into the underlying lava-flow, so forming another index of stratigraphic younging.

Neptunian Dykes Blocky lava-flows commonly have cracks and fissures extending downwards for varying depths into their central portions, while their upper surfaces are often covered by masses of jumbled-up blocks. Sediment can be trapped between these blocks, penetrating downwards to fill in the underlying cracks and fissures in the lava-flow itself, as shown in **81**. This sediment is often bedded, as if it was deposited by running water. It may also be altered by the residual heat of the lava-flow. The name given to such sedimentary infillings implies that the sediment was derived from the surface, Neptune being the sea-god of Greek mythology (see also **104**).

Figure **79** *Vesicular top of a lava-flow, overlain by conglomerate, formed by gas-bubbles rising through the lava-flow while it was still molten, to accumulate in a layer at its top. Crawton (NO 880796), Kincardineshire, Scotland.*

Figure **80** *Weathered bole affecting the scoriaceous top of a basic lava-flow, which in dipping at a moderate angle towards the right is overlain in the same direction by the base of the next flow to be erupted. Flodigarry (NG 467706), Isle of Skye, Scotland.*

Figure **81** *Neptunian dyke occupying a fissure in a vesicular lava-flow and filled with silty sediment that can be identified by its fine-grained nature and grey colour. Stannergate (NO 443310), Dundee, Scotland.*

Pillow Lavas

Pillow lavas are mostly formed by the eruption of basic lava-flows under submarine conditions. However, lava-flows can take on a pillowed structure if they are erupted into very wet mud, or even under an ice-sheet. **82** shows the typical appearance of such a lava-flow where it consists almost exclusively of a jumbled-up heap of lava-pillows. Adjacent pillows may be connected together by a neck of some sort. However, their separate identity can be inferred from the presence of a fine-grained or even glassy selvedge to each pillow. This is formed by the quenching of the molten lava as it came into contact with sea-water. Vesicles are often arranged in concentric zones close to the pillow margins, while jointing may be present, radiating from the centres of individual pillows. Any spaces between the pillows are commonly filled with a variety of material, including broken-up fragments of glassy lava, and secondary vein minerals, together with chert, limestone or shale.

It is evidently the submarine eruption of basic lava which generally leads to the formation of pillow lavas. As the lava issues forth, a glassy skin is formed on its surface, produced by the quenching effect of cold sea-water. Although capable of further distension as the lava continues to erupt, eventually this skin ruptures. This allows a bulbous protrusion of molten lava

to form. However, its surface is rapidly quenched as it comes into contact with the sea-water, forming a glassy skin to the protrusion itself. This prevents the enlargement of the protrusion beyond a certain size, although its centre may well remain molten for some time. A pillow-like mass of lava is formed as a result. This process appears to be repeated time and again as the lava continues to flow from its vent, eventually forming a lava-flow which consists almost entirely of lava pillows, at least towards its extremities.

Although lava pillows take on a variety of shapes, including elongate and flattened forms much like bolsters or sacks, it is the more ellipsoidal forms which present the greatest interest to the field geologist. The upper surface of each pillow occurs in the form of a low dome, as can be seen in **82**. The base of any pillow formed on top moulds itself to this shape, often bridging the slight gap between two adjacent pillows underneath, so forming a pillow of characteristic shape, as shown in **83**. The upper surface of such a pillow apears to become rigid before it is covered in its turn by more pillows. Any cross-section through a pillow lava often allows the direction of stratigraphic younging to be determined, as shown in **84**, simply by observing the contrast between cuspate bases and bun-shaped tops of individual pillows.

Figure 82 *Pillow lava showing how the whole mass of a lava-flow often appears to consist of pillow-like structures, forming separate masses of lava rarely more than one or two metres across, and often somewhat smaller. Newborough Warren (SH 391634), Anglesey, Wales.*

Figure 83 *Lava pillow showing a marked contrast between its cuspate base and a bun-shaped top, which provides an extremely useful index of stratigraphic younging in volcanic sequences formed by pillow lavas. Newborough Warren (SH 391634), Anglesey, Wales.*

Figure 84 *Pillow lava as seen in vertical cross-section. The slight elongation shown by the pillows indicates that the bedding dips very steeply to the left, while the shape of the individual pillows indicates that the lava-flow youngs to the right, so that it is slightly overturned. Newborough Warren (SH 391634), Anglesey, Wales.*

NATURE OF IGNEOUS INTRUSIONS

It is not perhaps immediately obvious why it should be thought that igneous intrusions, now consisting mostly of crystalline rocks, were originally emplaced in the molten form of magma. However, the crystalline textures characteristic of rocks occurring in the form of igneous intrusions rule out any possibility of a fragmental origin, whereby their constituents could have been introduced in a piecemeal fashion. Apart from hydrothermal solutions carrying mineral substances in an aqueous solution, from which they are deposited mostly in the form of mineral veins, a melt is apparently the only other agency capable of producing all the varied forms shown by igneous intrusions. A magmatic origin is further suggested by the circumstantial evidence that igneous rocks vary imperceptibly in character from intrusive rocks, very coarse in grain-size, to fine grained and even glassy rocks, characteristic of lava-flows. Since lava-flows can be seen to erupt in a molten form, it is reasonable to assume that even their coarse-grained equivalents crystallized out from a magma. In fact, many intrusions are formed by rocks almost identical in their lithology to lava-flows.

More evidence for a magmatic origin is shown by the presence of *chilled margins* to many intrusions of igneous rock. They are marked by a gradual decrease in grain-size towards an igneous contact over a distance to be measured in metres or centimetres, depending on the size of the intrusion. In some cases, a glassy selvedge is found along the very contact of the intrusion with its country-rocks. If the original intrusion chilled as magma at a high temperature against much cooler country-rocks, such features would be expected. In particular, the much more rapid cooling affecting the margins of the intrusion would result in the crystallization of a much finer-grained rock than that forming its centre, which must have lost heat much more slowly to its surroundings. Likewise, the presence of an aureole of thermal metamorphism around an igneous intrusion argues that the intrusion was itself a source of considerable heat, and therefore most likely to have been intruded as magma at a high temperature.

Major and Minor Intrusions

Although a great variety of names are applied to the forms shown by igneous intrusions, a broad distinction can be made between major and minor intrusions. *Major intrusions* are emplaced by a variety of mechanisms, generally at considerable depths within the earth's crust. They are of considerable size, measuring kilometres across even their smallest dimension. One consequence of their size is that they must have cooled down from a molten state over a long period of time. They are therefore formed of coarse-grained rocks, generally known as *plutonic* (Greek: *Pluto*, god of the underworld). Their country-rocks are affected by wide aureoles of contact metamorphism, while chilled margins of fine-grained rock may be lacking. Major intrusions often have a complex internal structure, formed by the repeated intrusion of magma, producing what are known as *internal contacts* between different parts of the intrusion as a whole.

Minor intrusions are much smaller than major intrusions, particularly in relation to their width or thickness, which is often less than a few tens of metres. Since they would then have cooled down much more quickly than a major intrusion, they are generally composed of rather fine-grained rocks which are known as *hypabyssal* (Greek: *hypo*, less than; *abussos*, a bottomless pit). Chilled margins are common to such intrusions, while they usually lack any aureole of thermal metamorphism, other than a narrow zone of contact alteration affecting the country-rocks in their immediate vicinity. Although minor intrusions are mostly formed by fine-grained rocks, *pegmatites* (Greek: *pegma*, a framework) are an exception to this general rule. These are extremely coarse-grained rocks with an igneous mineralogy. They often occur as minor intrusions, taking on their characteristically coarse-grained textures as the result of crystallization from a water-saturated residuum in the final stages of consolidation from granitic or other magmas.

Dilatational Intrusions

Since igneous intrusions do occur in juxtaposition with their country-rocks, they must have made a space for themselves in some way. This may simply occur by the country-rocks moving apart from one another on either side of a fracture, opening up a gap that can then be intruded by the magma if it is under sufficient pressure. It is then possible to match up the country-rocks across the intrusion, which has the parallel-sided form of a sheet-like body with its thickness limited in comparison with its other two dimensions. Since the space now occupied by the rocks of such an intrusion was formed by the dilatation of the country-rocks, it is known as a *dilatational intrusion*.

If the contacts of a dilatational intrusion are parallel to any structural features present in the country-rocks, such as the bedding, they are said to be *concordant*. If this is not the case, they are known as *discordant*. Sills and dykes (American spelling: dikes) are typical examples of dilatational intrusions, only differing from one another according to whether or not they are concordant to the bedding of their country-rocks, as shown in Drawing 2.

Dykes are parallel-sided intrusions of igneous rock that are markedly discordant to the bedding or any other structures in their country-rocks. This means in effect that most dykes are vertical or nearly so, intruded into flat-lying sediments. They often occur in the form of *dyke-swarms*, so that they are all roughly parallel to one another throughout a particular region. Dyke-swarms can be centred on an area of igneous activity, around which they may in fact be disposed in a radial fashion. Dykes can be taken as evidence of horizontal exten-

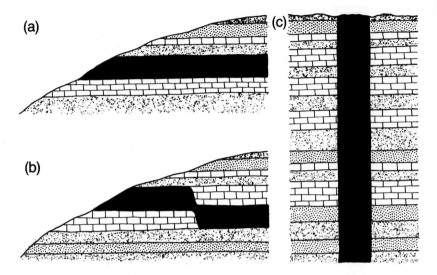

Drawing 2 *Vertical cross-sections through sheet-like intrusions:* (a) *concordant sill;* (b) *transgressive sill;* (c) *discordant dyke.* (*Roberts, Fig. 6.10,* Introduction to Geological Maps and Structures, *published by permission of Pergamon Press.*)

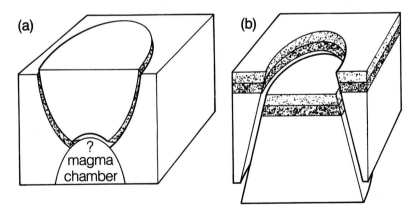

Drawing 3 *Block diagrams showing the typical form of* (a) *cone sheets and* (b) *ring dykes.* (*Roberts, Fig. 6.15,* Introduction to Geological Maps and Structures, *published by permission of Pergamon Press.*)

sion affecting the earth's crust at the time of their intrusion. *Sills* are sheet-like bodies of igneous rock which have intruded their sedimentary country-rocks along the bedding to form a concordant intrusion. Although they may be tilted subsequently along with their country-rocks, it seems most likely that the great majority of sills were intruded in a horizontal position, prior to these movements. Their intrusion implies that the magma was under sufficient pressure to lift its roof over a wide area, perhaps aided by the fact that it would have been denser than the overlying rocks actually forming its roof.

Other forms of dilatational intrusion include ring-dykes and cone-sheets, as shown in Drawing 3. They are both formed by the intrusion of magma into a system of curved fractures, which are often found to affect the country-rocks above an igneous complex. It is thought that these fractures dip very steeply outwards in the case of *ring-dykes*, allowing the central block of country-rocks to subside in a cauldron-like fashion, so forming the space that is now occupied by the ring-dyke. The fractures dip in towards the centre of the igneous complex at a moderate angle in the case of *cone-sheets*, so that the intrusion of magma under pressure into these fractures must cause uplift to affect the overlying country-rocks.

Volcanic Pipes   All these dilatational intrusions can be compared with the pipe-like conduits which act as feeders to the central eruptions of volcanoes. A distinction is usually made between *volcanic vents*, filled with pyroclastic rocks such as vent agglomerate and volcanic ash, and *volcanic plugs*, which consist entirely of igneous rocks, often in the form of a single intrusion. The space now occupied by such a volcanic plug is likely to have formed during an earlier stage of explosive activity, associated with the eruption of pyroclastic rocks. Volcanic vents filled with a mixture of pyroclastic rocks and igneous intrusions are more usually known as *volcanic necks*, although this term may be applied to any underground conduit that feeds a volcanic vent at the surface.

Nature of Intrusive Contacts

Igneous intrusions vary greatly in shape and size while they can be formed by a wide variety of different rock-types. However, they always have one feature in common. They are all intruded as magma into pre-existing rocks of the earth's crust, generally at some depth below the surface. This means that the rocks forming an igneous intrusion must be younger than their *country-rocks*, as the older rocks surrounding an igneous intrusion are known. Any structural features present in the country-rocks, now truncated by the igneous intrusion, must have formed prior to the phase of igneous activity which is represented by such an intrusive body.

Igneous intrusions have what is termed an *intrusive contact* with the surrounding country-rocks, against which they are juxtaposed as a very consequence of their intrusion. **85** shows a typical example. The light-coloured rocks in the foreground represent the margins of a granitic intrusion. The darker rocks in the background are the country-rocks to the intrusion. Originally deposited as shales, these rocks are now in the condition of hornfelsed slates. Their present character results from a combination of contact metamorphism, superimposed on a slaty cleavage formed by an earlier phase of earth-movements prior to the intrusion of the granite.

86 shows an example of a discordant contact between an igneous intrusion, formed by the rather dark rock on the left, and a finely-bedded quartzite. There is slight evidence of chilling along this contact, since the intrusive rock becomes finer-grained against the quartzite. Note how the contact cuts obliquely across the very fine laminations in the quartzite.

The contacts of igneous intrusions with their country-rocks are often more complex than the present examples might suggest. Even if it is marked by a sharply defined plane, this surface may show gradual or even abrupt changes in attitude as it is traced along its length. Tongues of igneous rock may penetrate the country-rocks away from the intrusion, forming veined or sheeted contacts on a larger scale.

Intrusive contacts can also be formed as a result of the explosive activity associated with the formation of pyroclastic rocks, as shown in **87**. If the walls of a volcanic vent had been so shattered by the volcanic explosions, the fractured and fragmented rocks would now occur in the form of *explosion breccias*. Such a rock passes into what is known as an *intrusion breccia* wherever it carries rock fragments derived from greater depths. By way of contrast, the fragments and blocks in a vent agglomerate mostly consist of volcanic rocks, rather than the surrounding country-rocks.

Figure **85** *Intrusive contact between the Godolphin Granite and its country-rocks, which passes just in front of the figure. The contact dips at 30° or thereabouts away from the camera, so that it forms more of a roof to the intrusion, than its walls. Rinsey Head (SW 593269), Cornwall, England.*

Figure **86** *Discordant contact made by an igneous intrusion with its country-rocks, formed by the pale-coloured quartzite on the right. Note how the igneous intrusion cuts obliquely across the fine laminations in the quartzite, so establishing that the intrusion is later in age. Buchaille Etive Mor (NN 227547), Glencoe, Scotland.*

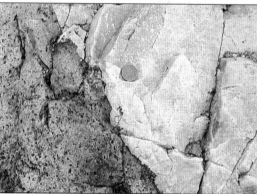

Figure **87** *Intrusive contact between a volcanic breccia, lacking any obvious structure apart from its fragmental nature, and the well-bedded quartzites in the foreground, forming the walls of a volcanic vent. Buchaille Etive Mor (NN 227547), Glencoe, Scotland.*

Dating of Igneous Intrusions

Chilled Margins It may be considered self-evident that igneous intrusions are always younger than the country-rocks which they intrude. However, it is not always easy to distinguish the rocks of an igneous intrusion from its country-rocks, particularly if the latter are igneous rather than sedimentary in character. Under these circumstances, evidence of chilling of one rock against another provides extremely useful evidence of relative age. An example is shown in **88**. The rocks on the right are dark shales with the bedding dipping very steeply to the left. They are juxtaposed against much lighter rocks on the left, representing the margins of an igneous intrusion formed by rather acid rocks. A chilled margin is developed to the intrusion, whereby the igneous rock becomes finer-grained as it is traced towards its contact with the country-rocks.

Older country-rocks are often altered in some way next to an igneous intrusion. **89** shows a thin vein, formed by a rather basic intrusion, cutting across an acid lava. Note how the country-rocks lying on either side of this vein are bleached for about a centimetre away from its contacts. Frequently, the country-rocks adjacent to an igneous intrusion become indurated in its immediate vicinity, while the wider effects of contact metamorphism leading to the formation of altered rocks such as hornfelses is also seen around igneous intrusion.

Cross-cutting Intrusions It is also possible to place igneous intrusions in chronological order wherever they can be seen to cut across one another. This then allows an intrusive history to be determined for the rocks in question. An example is shown in **90**, where two intrusions of igneous rock are present, one cutting across the other. The younger intrusion is granitic in composition, forming a horizontal sheet of pale-coloured rock, slightly iron-stained, which can be traced horizontally across the field of view. It evidently cuts across a vertical sheet of much darker rock, forming an older intrusion of more basic rock. This is intruded in its turn into even older country-rocks, formed by the light-coloured quartzite which occupies the remaining parts to the field of view. Note that it is rather difficult to trace the contacts of the granitic vein against the quartzitic country-rocks on either side of the more basic intrusion.

Further evidence of relative age may be obtained wherever inclusions of an older igneous rock are found as *xenoliths* (see **95**), completely enclosed by the rocks of a younger intrusion.

Figure 88 *Chilled margin to an igneous intrusion in contact with its country-rocks on the right. Note how this intrusion shows a decrease in grain-size towards its contact, as shown by the way that the slightly weathered faces of the joints cutting this intrusion become smoother in the same direction. Abereiddy (SM 801321), Dyfed, Wales. (Field of view c 1.2m)*

Figure 89 *Alteration affecting the igneous rocks of an older intrusion for about a centimetre along the sides of a younger vein of more basic rock, intruded along a narrow fracture in its country-rocks, Pennafort Gorge, Provence, France.*

Figure 90 *Cross-cutting relationships allowing igneous intrusions to be dated with respect to one another. A vein of granite forms a horizontal sheet, cutting across the older rocks of a more basic intrusion, itself intruding the even older quartzites which form its country-rocks. Kentallen (NN 011582), Argyll, Scotland.*

Contact Relationships of Sills

Since sills are found as igneous intrusions parallel to the bedding of the surrounding rocks, they have a form superficially similar to that shown by lava-flows, interbedded with sedimentary rocks. However, sills can be distinguished from lava-flows in the field, according to the nature of their contacts with the adjacent rocks.

Although sills generally have a contact which is concordant with the bedding of the underlying rocks, this is not always the case. **91** shows the well-known contact developed at the base of a dolerite sill, first described by James Hutton (1726–97) from Salisbury Craigs near Edinburgh in Scotland. The lower part of the sill is formed by the band of rather dark rock, which passes upwards into the lighter-weathering and much browner rock forming the bulk of the intrusion. The contact of the sill with the underlying sediments is seen immediately below this band of rather dark rock.

Although this contact can be seen to follow the bedding of the underlying rocks over much of its length, there are two points where it cuts across the bedding to form a discordant contact. Any sill which changes its stratigraphic level in such a step-like fashion is said to be *transgressive*. A closer view of this transgressive contact is shown in **92**. Such intrusive relationships are only found at the lower contact of an igneous body which forms a sill rather than a lava-flow.

Sills have upper contacts which are also intrusive into the overlying rocks, unlike lava-flows. **93** shows the discordant nature of the igneous sill where it can be seen to intrude the sedimentary rocks forming its roof. There is also a fine-grained selvedge present along this contact, forming a chilled margin to the intrusion. Tongues of igneous rock may also penetrate the overlying sediments from the main body of the intrusion. These features are not seen along the upper contacts of lava-flows, where blocks of the underlying lava may be completely surrounded by sediment, while it is the top of the lava-flow which is cut by tongues of sediment, penetrating downwards from the surface.

The country-rocks lying immediately below a sill often show some degree of contact alteration. Similar effects may also be seen below lava-flows, although they are usually not so marked. However, the rocks lying immediately above a sill are commonly altered in the same way by the effects of heat, whereas this is not seen in the case of lava-flows, except where dykes of fine-grained sediment penetrate cracks in the upper surface of a lava-flow while it still remains hot (see **81**).

Figure **91** *Transgressive contact between an igneous sill and its underlying country-rocks, cutting across the bedding of these rocks in two places. A closer view of this contact is provided by* **92**. *Hutton's Section, Salisbury Craigs (NT 270736), Edinburgh, Scotland. (Height of section c 4m)*

Figure **92** *Closer view of the transgressive contact shown in* **91**. *It can be seen that a wedge of magma has penetrated for a short distance along the bedding at the point where the intrusion changes level, tilting the overlying beds upwards so that they are now bent. Hutton's Section, Salisbury Craigs (NT 270736), Edinburgh, Scotland.*

Figure **93** *Discordant contact between an igneous sill and its overlying country-rocks. It can be seen how the rather massive rocks of this sill form an intrusive contact, cutting across the bedding of the overlying sediments at an oblique angle. Salisbury Craigs (NT 270728), Edinburgh, Scotland. (Field of view c 4m)*

Evidence for Magmatic Stoping

One way in which igneous intrusions can apparently make space for themselves involves a process known as *magmatic stoping*. This term is used in much the same sense as "overhead stoping" which is a method of underground mine-working. It implies that the magma penetrates its walls and roof along a series of fractures, detaching blocks of the country-rocks, which then sink into the magma. This process usually occurs on a small scale, when it is known as *piecemeal stoping*.

Occasionally, it is possible to see the initial stages in this process, frozen in time. 94 shows a dyke, which forms the vertical sheet of much darker rock, cutting across an earlier intrusion of fine-grained granite. It is a dilatational intrusion which has made a space for itself through the opening up of a whole series of parallel fractures, lying *en echelon* to one another, in its country-rocks. Where the intrusion has stepped sidewards directly from one fracture to another, the walls of the dyke can be matched up with each other across the intrusion. However, where the dyke has penetrated more than one fracture at the same level, as seen towards the top of the quarry, an intervening flake of the country-rocks, separates the two branches of the intrusion at this point. If the walls of the intrusion had moved farther apart, this block would have become completely isolated within the dyke. Such blocks of country-rock, completely surrounded by the rocks of an igneous intrusion, are known as *xenoliths* (Greek: *xenos*, a stranger). A xenolith, very nearly detached from its wall-rocks, can be seen at a lower level within the dyke.

Obviously, removing a piece of country-rock from the walls of an igneous intrusion provides a space for the magma to intrude, at least locally. Repeated time and again, it is a process which would gradually allow the magma to advance as an igneous intrusion into its country-rocks, at least in its upper levels. However, it is a process which does not create any space overall, since the space originally occupied by the xenolith is now occupied by the magma, and vice versa. Moreover, there is often a dearth of xenoliths at deeper levels of igneous intrusions, where they might be expected to accumulate.

This difficulty is usually avoided by assuming that the xenoliths become assimilated, perhaps even as a result of melting, as they sink into the magma, forming the rather irregular remnants as shown in 95. Such xenoliths are often intensely altered, so that their original lithology becomes difficult to establish. 96 shows how they may gradually lose their sharp outlines, eventually leaving diffuse patches that only differ somewhat in mineralogy from the rest of the rock.

Figure **94** *Dilatational intrusion forming an igneous dyke, cutting across an earlier intrusion of fine-grained granite. By matching up the walls of this intrusion, it can be seen how the original fracture-system opened up, so forming the space that it now occupies. Crarae Quarry (NR 996981), Argyll, Scotland. (Field of view c 25m)*

Figure **95** *Basic xenoliths forming discrete patches of rather darker rock within the Ballachulish Granite. Their irregular outlines suggest that these xenoliths have been assimilated to a considerable extent by the magma, almost as if they had started to melt. Ballachulish (NN 024595), Argyll, Scotland.*

Figure **96** *Xenolith of somewhat darker rock within the Shap Granite, showing rather diffuse margins. The larger feldspar crystals in this xenolith, similar to those found in the Shap Granite itself, must have grown in what was effectively a solid rock. Shap Granite Quarry (NY 556084), Cumbria, England.*

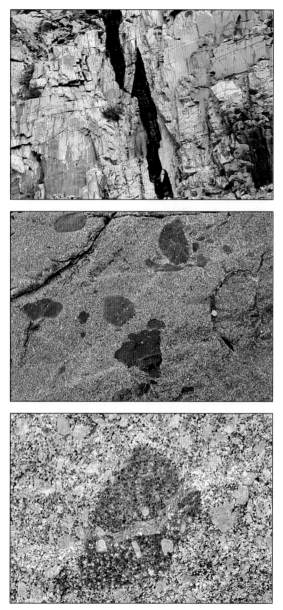

Columnar Joints and Sheet-joints

Columnar jointing is a characteristic feature of igneous rocks, formed in response to the contraction which occurs as they cool down and crystallize out from a molten state. It is most commonly found in lava-flows and igneous intrusions of a basic composition, typically affecting relatively fine-grained rocks, as shown in **97**. The columns themselves are formed by a series of very regular joints, forming a polygonal network. The joint-faces may show slight ridges on a small scale, lying at right angles to the length of the columns. These are thought to mark small steps in the propagation of the fracture, which now forms the joint-face itself. **98** illustrates the polygonal form of columnar jointing, as seen in cross-section. Although six-sided columns are the most common, columns with five or seven sides are also found. The flat or slightly curved surfaces cutting across the individual columns, exposing their cross-sections, are known as *cross-joints*.

Columnar jointing forms through the cooling of a sheet-like body of igneous rock. This results in a state of uniform contraction acting in all directions parallel to its contacts. The extension joints which form as a polygonal system in response to this contraction therefore occur at right angles to the *cooling surface*, which is formed by any contact to the igneous body (see **73**). This means that columnar jointing can be used to determine the attitude of an igneous contact with its country-rocks which is otherwise not exposed. For example, the columnar jointing found in a dyke is usually close to the horizontal, whereas it is often vertical in the case of lava-flows and sills. However, some degree of irregularity may occur, wherever the cooling is not entirely uniform.

Columnar jointing starts to form close to the contacts of the igneous body, propagating inwards to affect its interior as it cools. This can result in the presence of two sets of columar joints in a single lava-flow, often forming what is known as a *colonnade* of large and very regular columns, overlain by an *entablature* of smaller and less regular joints. This often gives a false impression that more than one flow is present.

Coarse-grained igneous rocks of a granitic composition may show *sheet-jointing*, which resembles columnar jointing in that it does not appear to have a tectonic origin. **99** shows a granitic rock containing the occasional xenolith, which is cut by a series of parallel joints, all dipping at a moderate angle to the left, parallel to the topography. Such joints become less abundant and more widely spaced with increasing depth, while they come to lie closer to the horizontal in the same direction. They are thought to form through the relief of pressure which occurs as erosion removes the burden formed by the weight of the overlying rocks.

Figure 97 *Columnar jointing forming the locality known as Samson's Ribs near Edinburgh. The columns formed by the jointing are very regular in form, extending for many metres along their lengths, despite their otherwise slender nature. Arthur's Seat (NT 274725), Edinburgh, Scotland.*

Figure 98 *Columnar joints in horizontal cross-section, showing the nature of the polygonal pattern typically formed by this type of jointing in igneous rocks. Uamh Oir (NG 372720), Isle of Skye, Scotland.*

Figure 99 *Sheet jointing in the Ballachulish Granite, forming a set of somewhat irregular joints, all roughly parallel to one another, dipping at a moderate angle to the left, and following the slope of the hillside as it descends into the sea. Ballachulish (NN 024595), Argyll, Scotland. (Field of view c 5m)*

EMPLACEMENT OF MAJOR INTRUSIONS

How major intrusions of igneous rock are emplaced in the pre-existing rocks of the earth's crust is often difficult to understand. It is not only the size of such intrusions that poses a problem, but also particular difficulties arise in considering the three-dimensional shapes of many intrusions. Indeed, these difficulties often start when attempting to establish exactly what shape is shown by a major intrusion as it is traced downwards from the surface.

Layered Intrusions

Many intrusions of basic and ultrabasic rocks developed on a large scale consist of distinct layers, varying from only a few centimetres to many hundreds of metres in thickness. Rhythmic banding commonly occurs on a small scale, forming layers in rich pyroxene and olivine near their base, passing upwards into rocks rich in feldspar towards their top. In addition, a lithological layering is often present on a much larger scale, forming layers that differ in mineral composition as they are traced upwards through the intrusion. Although layered intrusions can occur as discordant bodies, dyke-like or funnel-shaped in some cases, they mostly have the form of horizontal sheets, often many kilometres in thickness. In particular, these intrusions are typically found on a very large scale in the form of saucer-shaped *lopoliths* (Greek: *lopas*, a flat dish), that were formed by the down-sagging of the crust, presumably under the

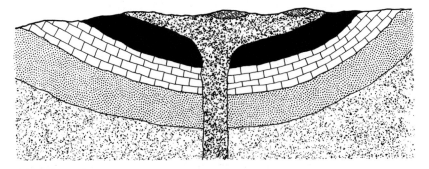

Drawing 4 *Vertical cross-section showing the inferred structure of a lopolith. (Roberts, Fig. 6.21,* Introduction to Geological Maps and Structures, *published by permission of Pergamon Press.)*

weight of the intrusion itself, as shown in Drawing 4. The layering has often been tilted away from its original position, close to the horizontal, by subsequent earth-movements.

Stocks and Batholiths

Major intrusions of granitic rocks often have the form of *batholiths* (Greek: *bathos*, depth). The contacts of such an intrusion dip steeply outwards as they are traced to greater depths, so that there is effectively no floor to the intrusion, as shown in Drawing 5. The outlines of granite batholiths, as shown on a geological map, provide a horizontal cross-section through these intrusions at ground level. Some batholiths have rather rounded outlines, while others are more elongate masses, lying parallel to the structural trends in the surrounding rocks.

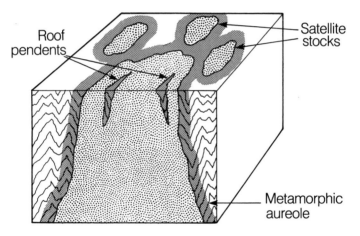

Drawing 5 *Block diagram showing the inferred form of granite batholiths. (Roberts, Fig. 6.23,* Introduction to Geological Maps and Structures, *published by permission of Pergamon Press.)*

Stocks are distinguished from batholiths wherever the intrusion is exposed over an area of less than 100 sq. km. While stocks may be formed by a single episode of intrusive activity, batholiths are usually found to consist of many separate intrusions, all closely related to one another. The internal contacts between these intrusions are often just marked by sharp but frequently rather subtle changes in mineral composition, without much evidence of one intrusion chilling against the other. In fact, the rocks of a slightly older intrusion may not have completely crystallized when they are penetrated by a later intrusion, giving rise to complex zones of hybridization between the two intrusions.

Forceful Intrusions

A magma which is so viscous that it cannot penetrate any thin cracks in its wall-rocks cannot then wedge them apart to form dilatational intrusions, nor otherwise allow piecemeal stoping to occur. However, it may have sufficient force to push aside its country-rocks to make space for itself. The simplest form is known as a *laccolith* (Greek: *lakkos*, a cistern). This is a concordant intrusion with a flat-lying floor, intruded into sedimentary rocks so that its roof now forms a dome, pushed upwards into an arch-like structure by the force of the magma. It differs from a sill in that the magma was too viscous to flow for any distance along the bedding before it started to solidify.

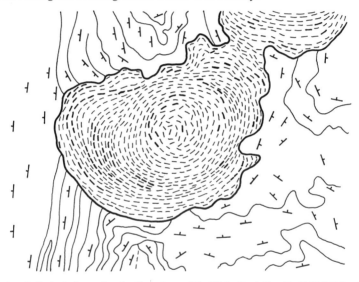

Drawing 6 *Geological map showing the outcrop of the White Creek Granite, British Columbia (after Buddington). The structure of the country-rocks is deflected close to the granite, while an internal foliation is found within the granite, parallel to its contacts. (Roberts, Fig. 6.26,* Intro-duction to Geological Maps and Structures, *published by permission of Pergamon Press.)*

Although laccoliths are generally developed on rather a small scale, so that they occur in the form of minor intrusions, the processes of forceful intrusion often operate on a much larger scale. It is under these circumstances that differences in density between the magma and its country-rocks can become important. For example, granitic magmas are often less dense than the country-rocks which they intrude, while they are also extremely viscous. The buoyancy of such a magma then allows it to intrude upwards, shouldering aside the country-rocks in a forceful fashion as it does so.

98

Forceful intrusions can be recognized from their effects on the surrounding rocks. Commonly, the bedding of the country-rocks is dragged upwards against the igneous contact, while folds and other structural features may be superimposed on these rocks, gradually becoming less intense so that they eventually disappear as they are traced away from the intrusion, as shown in Drawing 6. Although the central parts of a forceful intrusion may lack any structure, a foliation is often found towards it margins, passing outwards into a cleavage or schistosity which has been imprinted on the surrounding country-rocks as a result of their deformation. The country-rocks take on the characteristics of regionally metamorphosed rocks under these conditions.

Discordant Granites

While forceful intrusion provides a satisfactory explanation for the emplacement of many granites, it obviously cannot be applied to discordant intrusions which appear to have had absolutely no effect on their country-rocks. These intrusions, in particular, often have highly irregular and deeply embayed contacts, cutting across the structural features shown by the surrounding rocks in what appears to be an almost random fashion, as shown in Drawing 7.

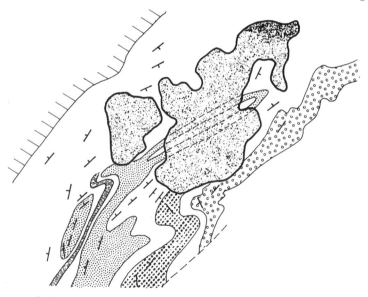

Drawing 7 *Geological map showing the outcrop of the Mullach nan Coirean Granite, Scotland, showing its discordant nature cutting across the structures in the country-rocks. (Roberts, Fig. 6.25,* Introduction to Geological Maps and Structures, *published by permission of Pergamon Press.)*

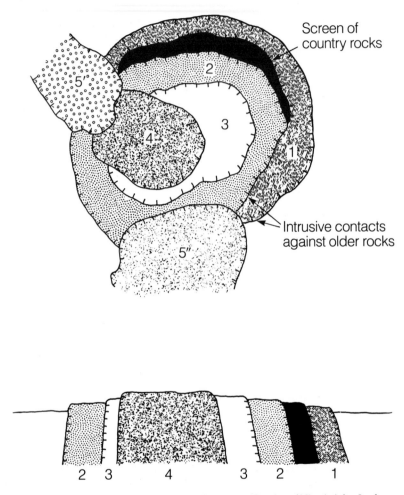

Drawing **8** *Ring complexes associated with the Younger Granites of Nigeria (after Jacobson, MacLeod and Black, and Turner). (Roberts, Fig. 6.17,* Introduction to Geological Maps and Structures, *published by permission of Pergamon Press.)*

Although reference is usually made to the process of piecemeal stoping to account for the emplacement of such granites, it has already been emphasized that piecemeal stoping is actually a process that does not create any space. It has therefore been argued that some bodies of an igneous-looking rock like granite can be formed by the complete replacement of their country-rocks, aided by the influx of chemically active fluids.

Such a process of replacement, taking place essentially in the solid, is known as *metasomatism* (Greek: *soma*, a body). It involves the removal of some chemical constituents from the original rock, and the introduction of others in their place, leading to the formation of what is effectively a new rock, entirely different in its mineralogy and chemical composition. However, geochemical evidence now suggests that, although this process may operate locally, converting the country-rocks in the immediate vicinity of an intrusion into igneous-looking rocks, it is unlikely to operate on a much larger scale.

It is therefore difficult to understand how discordant granites make room for themselves. Most likely, a combination of different processes are involved so that forceful intrusion at greater depths may be associated with piecemeal stoping at higher levels. Equally, it is possible that xenoliths are more easily assimilated into the magma than hitherto thought likely, so solving at least part of the space problem posed by discordant granites, particularly if any melting occurs.

Ring-Complexes

These consist of a whole series of concentric intrusions, all closely related to one another. The contacts between the various intrusions in a ring-complex are generally steeply inclined, so that they dip outwards at a high angle in most cases. This means that these intrusions appear to have the annular form typical of ring-dykes. However, it is usually found that the older intrusions occur around the outside of the complex, while they become progressively younger towards its centre, as shown in Drawing 8. It is then apparent that each intrusion makes an intrusive contact with the older rocks on its outside, while its inner margin is just formed by the rocks of a younger intrusion. The exact nature of any intrusive contact originally present along this inner margin is then simply a matter for conjecture, since it is now totally obscured by the later intrusion. Whether such an intrusion had the original form of a ring-dyke can only be determined if it is intrusive in older rocks along both its margins. These relationships are most likely to be preserved wherever the country-rocks are found as *screens* of older rock, separating the various intrusions from one another within a ring-complex.

Part iii
Unconformities and the Geological Record

Nature of Angular Unconformities

100 shows the classic example of an unconformity, exposed at Siccar Point, in Berwickshire, Scotland, where it was first discovered by James Hutton in 1788. The foreground is formed by vertical beds of greywackes and shales, which are Silurian in age. The unconformity is itself marked by the base of the overlying breccias and red sandstones, Upper Devonian in age, which dip at a low angle towards the left.

Angular unconformities are produced as a result of earth-movements, accompanied in particular by uplift and erosion, which effect an older sequence of sedimentary rocks, prior to the deposition of a younger sequence. An unconformity therefore forms a structural break between two sedimentary sequences, differing from one another in stratigraphic age. The structural discordance between the two sequences results from the fact that it is only the underlying rocks which are affected by the earth-movements that occurred during this interval of geological time. These rocks therefore no longer retain an attitude, close to the horizontal, with which they were first deposited.

The uplift which typically accompanies such earth-movements then allows a surface of erosion to cut across the structures now present in the underlying rocks. It is upon this surface that the overlying rocks are subsequently deposited, with little or no dip, once the deposition of sedimentary material is resumed. The unconformity is therefore marked by flat-lying beds of the younger sequence, resting unconformably on a surface of erosion which cuts across and truncates the structures now only found in the underlying rocks.

101 is a close-up view of the unconformity at Siccar Point. The underlying beds are greywackes, dipping steeply to the right. They are truncated by an irregular surface of erosion, on top of which the overlying red sandstones have been laid down. The bedding of these sandstones dips at a much lower angle towards the left. Angular fragments of the underlying greywackes are seen close to their source, embedded in these sandstones. This can be taken as further evidence for the erosion undergone by the Silurian rocks, prior to the deposition of the Upper Devonian sandstones.

The effect of subsequent earth-movements on an unconformity is shown in **102**. The unconformity itself occurs immediately below the prominent bed of sandstone, dipping at a moderate angle to the right. The underlying rocks are slates in which the bedding is close to the vertical. Their original dip can be determined if the dip affecting the unconformity itself is removed. Note how the rocks truncated by the unconformity are always older than the rocks lying stratigraphically above the unconformity. This important index of stratigraphic order can be used to determine the relative ages of stratigraphic sequences, separated from one another by an unconformity, even if it has subsequently been inverted.

Figure **100** *Exposed angular unconformity forming an unconformable contact between vertically-dipping greywackes and shales in the foreground, and the overlying breccias and red sandstones of Upper Devonian age, dipping at a low angle towards the left. Siccar Point (NT 813710), Berwickshire, Scotland.*

Figure **101** *Closer view of the angular unconformity at Siccar Point, as shown in the previous photograph. The unconformity itself forms an irregular surface of erosion, while the red sandstones lying on top of this surface contain angular fragments derived by erosion of the underlying rocks. Siccar Point (NT 813710), Berwickshire, Scotland.*

Figure **102** *Angular unconformity affected by subsequent earth-movements so that it now dips at a moderately steep angle towards the right. The original dip of the rocks lying stratigraphically below the unconformity can be determined if the dip affecting the unconformity itself is removed. Irede, Leon Province, Spain.*

USE OF UNCONFORMITIES IN STRATIGRAPHIC DATING

Since an unconformity represents a break in the stratigraphic record, it can be used to date any event that occurred during the corresponding interval of geological time, as long as this event left its imprint on the structure of the underlying rocks. However, it should first be emphasized that a part of the geological record is always lost as a result of the erosion which accompanies the development of an unconformity. Thus, deposition of the original sequence could have continued long after even the youngest rocks now preserved below an unconformity were laid down, nearly to the time when sedimentation was resumed, leading to the deposition of the overlying rocks. This means in effect that the longer gap represented by a particular unconformity, the greater the uncertainty there is concerning the precise dating of any intervening events.

Earth-movements

Since sedimentary rocks are nearly always laid down very close to the horizontal, any structural discordance present across an unconformity must reflect the nature of the earth-movements that occurred prior to the deposition of the overlying rocks. Apart from faulting, tilting or folding can affect the underlying rocks as the result of such movements, as shown in Drawing 9.

If only *tilting* has occurred, rocks lying immediately above the unconformity come into contact with younger and younger horizons in the underlying sequence as it is traced in the same direction at its dip. This relationship is known as *overstep*. It results from the tilting of the underlying rocks away from the horizontal position in which they were first deposited. This then allows erosion to cut a bevelled surface across the bedding of the underlying strata, prior to the deposition of the overlying rocks. Since the presence of an unconformity can still be recognized, with even the slightest degree of structural discordance between the two sequences, it can be argued that the depositional surface needs only to be distorted by a very small amount for erosion to take the place of deposition. This implies that sedimentary rocks are deposited with an initial dip that is extremely close to the horizontal under most circumstances.

If *folding* occurs during the interval of geological time represented by an unconformity, this fact will be reflected in the nature of the overstep shown by the overlying rocks. In particular, any reversals in the dip of the underlying rocks, occurring as a result of such movements, will be marked by first older then younger rocks coming into contact with the unconformity as it is traced in a particular direction across the folds present in the underlying rocks. The nature of the folding which affected these rocks can be determined from this evidence, even if the unconformity has itself been folded at a later date.

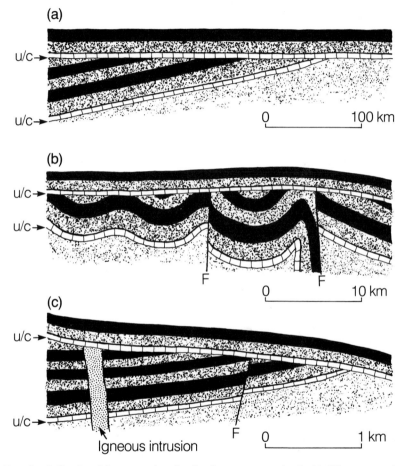

Drawing 9 *Stratigraphic cross-sections showing the overstep associated with different types of angular unconformity (U/C), according to the nature of the earth-movements affecting the underlying rocks: (a) tilting; (b) folding; (c) faulting. Each cross-section shows the lower stratigraphic sequence to rest unconformably on basement rocks. Note the differences in horizontal scale between the various cross-sections. (Roberts, Fig. 7.3,* Introduction to Geological Maps and Structures, *published by permission of Pergamon Press.)*

Faulting

Any fault seen to affect only the rocks lying below an unconformity must have formed prior to the deposition of the overlying rocks, wherever it is truncated by the unconformity itself. However, faults often undergo repeated movements, so that a fault formed during the interval of geological time now represented by an unconformity may subsequently be rejuvenated, so that it comes to affect the overlying rocks as well. Such a fault is often difficult to distinguish from one where all the movements occurred after the rocks lying above the unconformity were deposited. A clue may be provided wherever the displacement across the fault is much greater where it cuts the rocks lying below the unconformity, in comparison with its effects on the overlying rocks.

Igneous Intrusion

Unconformities are extremely important in dating the intrusion of igneous rocks, often allowing an upper limit to be placed on their stratigraphic age. It has already been emphasized that igneous intrusions must be younger than the country-rocks which they intrude. This means that a lower limit can be placed on the age of an igneous intrusion wherever its country-rocks can be dated stratigraphically. However, there is a difficulty, since the intrusion of an igneous rock must always occur at some depth below the earth's surface, so that it may be much younger than its country-rocks. In fact, if its country-rocks occur as a stratigraphic sequence, it is commonly assumed that any intrusion cutting these rocks is entirely younger than the stratigraphic sequence as a whole, not just the country-rocks in its immediate vicinity. This means in effect that the igneous intrusion is assumed to date from that interval of geological time, which is now represented by the presence of an unconformity, separating the intruded country-rocks from an overlying sequence of sedimentary rocks, lacking any such intrusions.

The only exception that can be made to this very general assumption concerns the dating of igneous intrusions which are obviously related in some way to the eruption of volcanic rocks, themselves preserved as part of the stratigraphic sequence in question. Such an intrusion can then be taken as contemporaneous with this volcanic activity, so allowing it to be dated stratigraphically.

The upper limit to the age of an igneous intrusion can be determined wherever it is overstepped by sedimentary rocks lying above an unconformity. It must then be older than these rocks. However, it is often found that, although igneous intrusions penetrate the rocks lying below an unconformity, they are nowhere found in contact with the overlying rocks. The evidence that they are older than these rocks then becomes rather circumstantial unless it can be shown that the rocks overlying the unconformity contain rock fragments or mineral grains clearly derived from the weathering and erosion of a particular intrusion within the underlying rocks.

Synclinal fold affecting the sedimentary rocks of the Esla Nappe in the Cantabrian Mountains of Northern Spain.

Unconformities as Erosion Surfaces

The structural discordance shown by an unconformity clearly indicates that it represents a surface of pronounced erosion, affecting the underlying rocks. 103 shows such an erosion surface, which marks the unconformable contact between a Carboniferous sandstone and an overlying breccia of Permian age, from the North of England. The breccia carries fragments of the underlying Carboniferous rocks, showing that it is younger in age. The presence of rock fragments in a sedimentary sequence, which are clearly derived from a source that can be identified as an older group of rocks, provides a very useful index of stratigraphic order, particularly on a regional scale. It goes almost without saying that the rock fragments need to be sufficiently distinct for their source to be identified without any doubt.

Another result of erosion acting along an unconformity may be seen in the opening-up of fissures in the underlying rocks. If these are then filled with sediment from the surface, they form *Neptunian dykes* (see also **81**). A typical example is shown in **104**. What may be termed the country-rocks are formed by gneisses, varying somewhat in mineral composition. The banding developed within these gneisses appears vertical in the photograph. They are cut by a thin vein of sedimentary material, consisting of a sandy mudstone, chocolate brown in colour. Angular fragments of the gneiss, detached from the walls of the fissure, now lie isolated within the Neptunian dyke. Field evidence shows that such fissuring becomes more pronounced upwards until it eventually passes into what is effectively a basal breccia, developed along the unconformity between the gneisses and an overlying sequence of sedimentary rocks.

The rocks lying below an unconformity may also show the effects of weathering. Commonly, they are stained by the altera-tion of iron-rich minerals in a way that does not affect the overlying rocks. Even although soil-profiles are rarely preserved below an unconformity, there is frequently some evidence to show that the underlying rocks have been affected by physical weathering. As shown in **105**, this is often marked by the break-up of the underlying rocks, which are then incorporated as fragments into the overlying sequence. The underlying rocks in the present case are high-grade metamorphic rocks, cut by granite veins. They pass upwards into a rubbly zone, lying immediately above what can be taken as the surface of unconformity, dipping at a low angle to the left. Traced upwards, this passes into a basal breccia, carrying granitic fragments clearly derived from the underlying rocks.

Overlap and Buried Landscapes The degree of topographic relief shown by an unconformity, can be determined by examining the disposition of the overlying strata. **106** shows an unconformity between flat-lying Carboniferous Limestone and underlying slates of Silurian age in the North of England. The bedding of the overlying limestones is parallel to the surface of erosion, which forms the unconformity at their base, showing that it had been reduced to the horizontal, prior to its deposition. This is typical of surfaces of marine erosion, formed as a result of transgressions by the sea across the land.

107 shows for comparison another view of the unconformity at Siccar Point (see **100**). The Upper Devonian sandstones and breccias in the foreground are clearly banked against low cliffs, formed by the underlying Silurian rocks, which can be seen in the background. These sandstones and breccias therefore occur as beds which wedge out against the Silurian rocks, while the unconformity itself occurs as a surface of erosion which shows a considerable degree of topographic relief.

Figure **103** *Angular unconformity showing a well-developed surface of erosion cut into the underlying Carboniferous sandstone, and overlain by a breccia of Permian age, which fills in the hollows formed in this surface as the result of differential erosion. Saltom Bay (NX 958158), Cumbria, England.*

Figure **104** *Neptunian dyke of sandy mudstone, chocolate brown in colour, cutting discordantly across banded gneisses and lying below a major unconformity between these gneisses and an overlying series of red sandstones and breccias. Clachtoll (NC 039268), Sutherland, Scotland.*

Figure **105** *Angular unconformity between high-grade metamorphic rocks forming a basement underneath, cut by granitic veins, and an overlying series of breccias and sandstones. Note how physical weathering has resulted in the break-up of the rocks lying immediately below the unconformity. Portskerra (NC 877666), Sutherland, Scotland.*

This relationship usually occurs immediately above an unconformity wherever one bed overlaps another to cover a wider area of deposition so that it wedges out against a surface of older rocks. It is therefore known as *overlap*.

Overlap typically occurs wherever the original surface of erosion was inclined, even slightly, away from the horizontal. The sedimentary rocks deposited on top of this surface would have shown an initial dip closer to the horizontal than that now exhibited by the unconformity itself. This is typical of unconformities developed under continental conditions, wherever erosion was not able to reduce the former topography to a horizontal surface before the deposition of sedimentary material was resumed. A buried landscape is formed as a result, while the overlying sediments often accumulate in the form of scree deposits or alluvial fans.

An example of a buried topography is shown in **108**. It is formed by high-grade metamorphic rocks, intruded by granitic veins, which are exposed as two knolls on either side of the bay. These rocks, pink in colour and lacking any bedding, are overlain unconformably by breccias and sandstones, forming the cliffs to the bay. The contact descends to the high-water mark as it is traced across the bay between these two knolls, showing a relief of a few metres in height. The overlying sediments is draped across the underlying rocks to form what is known as a *compaction fold*. Such a structure is partly an original feature, reflecting the development of initial dips away from the knolls formed by the underlying rocks, which results in the overlap shown by the sedimentary rocks lying on the flanks of the knolls. However, it is probably accentuated by the effects of *differential compaction*, reflecting the greater thickness of sedimentary rocks which were deposited in the depression between the two knolls.

109 shows an extreme form of such a buried landscape. The conglomerate with well-rounded boulders is Middle Devonian in age. It is banked against a near-vertical contact of much older rocks, as seen on the right. The contact represents an unconformity with an extreme degree of topographic relief, evidently marked by the presence of cliff-like features in the ancient landscape which it now represents.

Figure **106** *Angular unconformity between steeply-dipping slates of Silurian age, lying below the unconformity, and horizontal Carboniferous Limestone. The bedding of the overlying rocks is parallel to the surface of erosion forming the unconformity itself. Horton-in-Ribblesdale (SO 800702), Yorkshire, England. (Height of section c 12m)*

Figure **107** *Overlap associated with Hutton's Unconformity at Siccar Point in Berwickshire, Scotland, showing how the overlying rocks come into contact with the underlying greywackes and shales across what is effectively a very ancient landscape. Siccar Point (NT 813710), Berwickshire, Scotland.*

Figure **108** *Angular unconformity between the Old Red Sandstone of Middle Devonian age and an underlying basement of high-grade metamorphic rocks. The surface of erosion forming the unconformity itself represents a buried landscape, cut into the older rocks. Portskerra (NC 877666), Sutherland, Scotland.*

Figure **109** *Buried landscape of Middle Devonian age, showing a conglomerate containing well-rounded and very large boulders lying against much older rocks across a near-vertical contact, which forms a cliff-like feature in this ancient landscape. Quarry Head (NJ 898651), Buchan, Scotland. (Field of view c 3m)*

UNCONFORMITIES AND MOUNTAIN-BUILDING

Once unconformities were first recognized by James Hutton in the late 18th century, they formed the essential foundation for understanding the inner dynamics of the earth. Indeed, the development of geology as a historical science, concerned with deciphering a chronology of geological events from the record now preserved by rocks of the earth's crust, could only proceed from this degree of understanding. The fundamental importance attached to unconformities stems firstly from the fact that it is often possible to trace unconformities far and wide, so that they can be identified as geological structures which are present on a regional scale. Secondly, they are usually found to occur in any one region at a particular horizon within the stratigraphic column. Admittedly, the rocks lying below an unconformity can vary widely in age, but this is simply a consequence of the erosion which has affected these rocks prior to the deposition of the overlying strata. Likewise, the self-same rocks occurring immediately above an unconformity may also be found to vary in stratigraphic age wherever the deposition of sedimentary rocks has slowly encroached upon an area previously undergoing uplift and erosion, to form an unconformity marked by overlap of the younger strata.

Despite all these complexities, the gap in stratigraphic age between the *youngest* of the older rocks lying below an unconformity, and the *oldest* of the younger rocks occurring above the unconformity, often only represents quite a short interval of geological time. This is particularly true when it is compared with the length of geological time that is otherwise marked by the deposition of sedimentary rocks. This means that much of the geological record is concerned with the deposition of sedimentary rocks, punctuated by quite short-lived episodes when these rocks are affected by folding and faulting, accompanied by igneous intrusion and regional metamorphism in many cases. It is these episodes, closely associated with the processes of *mountain-building*, which are now represented by angular unconformities, occurring on a regional scale between sedimentary sequences that otherwise differ in stratigraphic age from one another. It should be stressed, however, that this is a somewhat simplistic view of geological history if it gives the impression that earth-movements only happened once in a while. Instead, even though earth-movements accompany the deposition of sedimentary rocks, it can be argued that they wax and wane in intensity over the course of geological time, giving rise to structural features like angular unconformities, which mark the climax of such movements.

The concept of mountain-building is not just applied in geology to the origin of mountains, simply as features of the landscape. Instead, it refers to the intense folding and faulting, characteristic of the rocks which are subsequently uplifted along broad zones of the earth's crust to form mountain chains like the Alps or the Himalayas. This distinction is maintained by the use of *orogeny* (Greek: *oros*, a mountain) as a special term to describe the various processes involved in mountain-building. Acting together, these processes result in the structural features now shown by *orogenic belts*, as the elongate zones affected by mountain-building are known. Some orogenic belts show that folding rather than faulting has been predominant in their structural evolution, and vice versa, while other belts show the effects of widespread deformation and regional metamorphism, perhaps accompanied by the extensive intrusion of igneous rocks.

Many orogenic belts show sedimentary rocks to be affected by mountain-building. Moreover, these sedimentary rocks often form very thick sequences, frequently interbedded with volcanic rocks, which were deposited over considerable lengths of geological time. Such sequences may be seen to rest unconformably on even older rocks. These rocks then occur as the foundations to the orogenic belt in question, so that they form what is known as its *basement*, while the sedimentary rocks lying within the orogenic belt are found as a *cover* to this basement. The rocks forming a basement to an orogenic belt are commonly just the eroded remnants of an earlier orogeny, or they are themselves underlain by such a basement.

Orogenic belts often have a geological history that is extremely complex in its details. Although the early stages in this history are usually marked by the deposition of sedimentary rocks, often accompanied by volcanic activity, they may well be interrupted at various times by earth-movements of varying intensity, leading to the development of unconformities between different parts of the stratigraphic sequence, at least locally. However, orogeny nearly always ends eventually with the uplifting of the rocks lying within the orogenic belt as a whole. After erosion, it is these rocks which form a basement upon which sedimentary rocks can be deposited unconformably, so beginning the next cycle in the geological evolution of the earth's crust.

Part IV
Mountain-building and the Tectonic Record

Sedimentary rocks affected by folding and thrusting as a result of earth-movements. Broad Haven, Wales.

118

Role of Plate Tectonics

Mountain-building, the intrusion of igneous rocks and regional metamorphism, together with volcanic activity, all imply that the earth has within itself a source of thermal energy. Before the nature of radioactivity was discovered in the late 19th century, this heat was thought to be primeval in origin, dating back to the formation of the earth itself. This led to the *contraction theory* of mountain-building, whereby the earth's crust was thrown into horizontal compression as the earth's interior cooled down, perhaps from a molten state, so accounting for the folding and faulting now seen within mountain belts. However, the discovery of radioactivity provided another source of thermal energy which, although it has become progressively less important with the passing of geological time, still drives the heat engine formed by the earth itself. In particular, it is this energy which is now thought to provide the driving force for the mechanisms of *plate tectonics* (Greek: *tekton*, a builder).

Plate tectonics has the merit of being a theory that is extremely simple in its outlines. It first recognizes that the earth's surface is divided up into a small number of so-called plates, each measuring up to a few thousand kilometres across, and extending as relatively solid rock to a depth of a few hundred kilometres at the very most. These plates may be capped by oceanic crust, only a few kilometres in thickness, which forms the floor of the deep oceans beyond the shallow seas around the continents. Alternatively, continental crust may be present, consisting of much lighter rocks with an average thickness of some 35km. These rocks rest on top of much denser rocks which are found beneath both continent and ocean, forming the foundations of the plates themselves. These denser rocks are underlain in their turn by much weaker material, close to its melting point, which allows the plates to move about the earth's surface with respect to one another.

How this occurs constitutes the other important element in the theory of plate tectonics. In fact, the plates can only move apart from one another where oceanic crust is produced as a result of sea-floor spreading. This forms what is known as a *constructive boundary* between two plates, forming a part of the mid-oceanic ridge system which encircles the globe. The oceanic crust produced in this way is eventually destroyed along a *destructive boundary*, which is formed wherever two adjacent plates move towards one another. Oceanic crust on one side of such a boundary descends below the other plate along an inclined surface known as a *subduction zone*, undergoing a phase change into much denser rocks as it does so, and forming an oceanic trench at the earth's surface as a result. Finally, a *conservative boundary* may be formed by what is otherwise known as a *transform fault*, wherever two adjacent plates slide past one another in a horizontal direction so that oceanic crust is neither created nor destroyed.

Plate Tectonics and Mountain-building

The processes of mountain-building are concentrated along the destructive boundaries, formed by the subduction of oceanic crust below an adjacent plate. These boundaries fall into two distinct types. If oceanic crust undergoes subduction against an adjacent plate capped by continental crust, a cordilleran type of destructive boundary is formed as a result, like the Andes. Alternatively, if the subduction of oceanic crust occurs beneath a plate which is itself capped by oceanic crust, this leads to the evolution of island arcs, as found in Japan. Both these types of destructive boundary are marked by abundant earthquakes and much igneous activity, while their geological evolution commonly leads to the development of structural features that are typical of orogenic belts as a whole. Even so, these features are often emphasized by the effects of continental collision, which eventually occurs as the final stage in the evolution of such a destructive boundary.

Continental crust differs in a number of ways from oceanic crust, which consists largely of basaltic rocks. Firstly, it is much closer to granite in composition, so that it is formed by much less dense rocks than oceanic crust. This is why the continental land-surface stands much higher than the ocean floors, while the continents extend to much greater depths in comparison with the thickness of oceanic crust. Secondly, continental crust cannot be converted into denser rocks as a result of phase changes occurring under such temperatures and confining pressures as exist relatively close to the earth's surface. It differs in this respect from the basaltic rocks of the oceanic crust, which can be altered to form a much denser rock known as *eclogite*. This means in effect that the continents are buoyant masses, made up of rocks with a composition that cannot be subducted into the much denser rocks occurring at greater depths within the earth. Accordingly, it is the continents which preserve much of the geological record, since they consist of rocks dating back to the earliest Precambrian, while the oceans are only underlain by much younger rocks, which have so far escaped the effects of subduction.

Continental crust therefore only plays a passive role in the movements that come under the heading of plate tectonics. Continents can be split apart by rifting, separating into fragments which drift apart as oceanic crust is formed along the intervening boundary. Carried along in a "piggy-back" fashion by the plate movements, these continental masses eventually come into contact with a destructive boundary, where subduction of oceanic crust was previously taking place. Unable to undergo subduction, this buoyant mass of continental crust collides with the rocks forming the other plate. It is these collisions which are preserved in the geological record by important phases of mountain-building, since they effectively put a stop to any further subduction. This implies that most orogenic belts have a *suture* within their boundaries, marking the junction between the two continental plates which came into contact with one another to form the orogenic belt itself.

Plate Movements and Tectonic Forces

The role of plate tectonics in mountain-building provides a background to any discussion concerning the nature of the tectonic forces acting on the rocks of the earth's crust. It is clear that plate movements must occur in response to forces, perhaps generated by convection currents deep within the earth, which are exerted in some way on the actual boundaries of the individual plates.

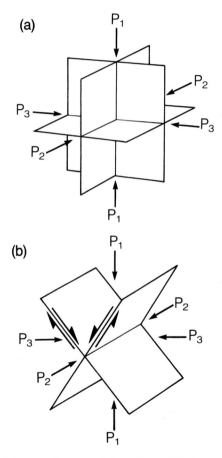

Drawing 10 *Internal stresses under mechanical equilibrium:* (a) *principal stresses acting on three planes lying at right angles to one another;* (b) *planes of maximum shearing stress at 45° to the direction of maximum compression.* P_1 = *direction of maximum compression;* P_2 = *direction of intermediate compression;* P_3 = *direction of minimum compression.*

However, since the movements undergone by these plates will be constrained according to the nature of their boundaries, internal forces will be set up, affecting the rocks lying within these plates, but concentrated in particular along their boundaries. Other forces will be generated wherever any sort of gravitational instability is produced as a result of geological processes acting on the rocks of the earth's crust, leading to what is known as *gravity tectonics*. These various forces if they exceed the long-term strength of the rocks under their influence, cause permanent deformation to occur. The geological structures to be described in the second half of this book are formed as a result.

The detailed analysis of how geological structures are formed as a result of deformation depends on understanding the nature of the internal forces acting on the rocks of the earth's crust. If we are just dealing with conditions of mechanical equilibrium, as implied by Newton's Second Law, any system of internal forces acting at a particular point can be considered in terms of *stress*. This simply measures the intensity of any force by taking the area over which it acts into account. There are then only three directions passing through the point, lying mutually at right angles to one another, in which the stresses are just compressions of extensions, lacking any component of shearing force. They are known as *principal stresses*. In fact, a direction of maximum compression generally lies at right angles to a direction of minimum compression, while the third direction lies at right angles to the plane containing the other two directions, as shown in Drawing 10(a). The direction of minimum compression may become a tension close to the surface, where the confining pressure produced by the weight of the overlying rocks has a sufficiently low value. However, it is common enough in structural geology to refer to all compressions below a mean value as given by the confining pressure as extensions.

All these compressions or extensions act on planes lying mutually at right angles to one another, while all the other planes passing through the point in question are affected by shearing forces as well. These forces in acting on a particular plane generate what are known as *shear stresses*. There are two planes of maximum shear stress, lying at 45° to the directions of maximum and minimum compression, while they intersect one another in the third direction, at right angles to the other two directions, as shown in Drawing 10(b). The shear stress acting on these planes has a maximum value equal to half the difference between the maximum and minimum compressions. It is this stress difference which causes permanent deformation in rocks, once it exceeds a value corresponding to the long-term strength of the rock itself.

Brittle folding and faulting of a layered rock, itself the result of ductile deformation above a major thrust-plane. Knockan Craig, Scotland.

Modes of Mechanical Behaviour

How a rock deforms permanently in response to any stress difference which exceeds its long-term strength depends on a number of factors. However, a broad distinction can first be made between *brittle fracture*, in which a rock breaks apart into a number of separate fragments, and *ductile deformation*, where it changes its shape without any loss in the physical cohesion of the rock itself. There is in fact a complete gradation in mechanical behaviour between these two extremes, marked by an increasing ductility of the rock as the physical conditions change.

110 shows a boulder from a deformed conglomerate, which illustrates both the effects of brittle fracture and ductile deformation. It is a reasonable assumption that this boulder would have had a much more rounded form when it was first deposited. Its present elongation, which is shared by other boulders in the same conglomerate, must be a result of deformation. It can be seen that this has been achieved partly as a result of brittle fracture, which has produced a network of intersecting fractures. Some of these fractures have opened up as gashes, so contributing to the elongation shown by the boulder itself, while shearing movements have occurred on other fractures, with the same effect. However, even if all these effects contributing to the present shape of the boulder were removed, it would still show a pronounced degree of elongation, which can only be the result of ductile deformation.

What sort of mechanical behaviour is shown by a particular rock once its long-term strength is exceeded depends to a considerable extent on its lithology. If the physical conditions are otherwise the same, crystalline rocks are the least ductile, exhibiting rather brittle behaviour under most circumstances, followed in their turn by quartz-rich sedimentary rocks, limestones and dolomites, shales and mudstones, and finally salt deposits, listed in order of increasing ductility. **111** provides an example, taken from a deformed conglomerate, in which boulders differing in their lithology have not all responded to the deformation in the same way. In particular, there is a large boulder of quartzite forming the centre of the field of view, which is hardly deformed at all, while the many smaller boulders of brown-weathering dolomite show a considerable degree of deformation.

Considering only one particular lithology, rocks always become more ductile with depth as the confining pressure increases in this direction. This tendency is reinforced by the rise in temperature which occurs in the same direction. Furthermore, the extremely slow rate at which deformation takes place in the earth's crust greatly enhances the ductility of the rocks so affected, while it has a very considerable effect in reducing the long-term strength below which such rocks will not deform in a ductile fashion. **112** shows how a conglomerate can show an extreme degree of deformation under these circumstances.

Figure **110** *Boulder from a deformed conglomerate, showing how it has become more elongate in shape than originally would have been the case as a response to the combined effects of brittle fracture and ductile deformation. Oughty Crag (L 930757), Connemara, Ireland.*

Figure **111** *Deformed conglomerate, showing marked differences in shape between boulders differing from one another in lithology. The small boulders of brown-weathering dolomite are strongly deformed, while the much larger boulder of quartzite in the centre of the field of view is hardly deformed at all. Skerrols (NR 350646), Isle of Islay, Scotland.*

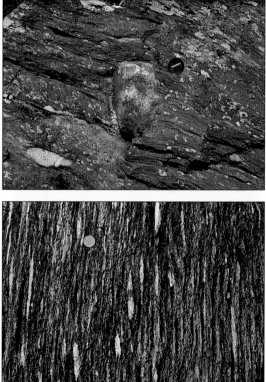

Figure **112** *Deformed conglomerate showing quite extreme amounts of ductile deformation. Note how the boulders whatever their lithology are all stretched out in the same direction, so that the rock as a whole has a strongly defined fabric as a result. Cuddie's Point (NG 913378), Wester Ross, Scotland.*

Nature of the Structural Record

Mechanical Effects of Pore Pressure

There is one factor that opposes any tendency towards greater ductility in rocks undergoing deformation. This is the effect of pore pressure on the mechanical behaviour of rocks saturated with pore fluids in the spaces between their grains. Under equilibrium conditions, this pressure will be hydrostatic in nature, directly proportional to the density of the pore fluid at any particular depth below the surface. Given the average density of sedimentary rocks, the hydrostatic pressure will be somewhat less than half the confining pressure, corresponding to the weight of the overlying rocks.

However, sedimentary rocks undergoing compaction do not form a solid framework of mineral grains, allowing the free movement of any pore fluids which is needed to maintain such a hydrostatic pressure. Instead, this framework tends to collapse in a progressive fashion as the confining pressure increases as the deposition of sedimentary rocks continues at the surface. This compresses the pore fluids in the spaces between the mineral grains, causing a rise in the pore pressure, often to a value close to the confining pressure. These anomalously high pore pressures are transitory in nature, eventually declining in amount to a value corresponding to the hydrostatic pressure as the pore fluids are gradually expelled from the sedimentary rock as the result of its further compaction. They are typically developed within sedimentary sequences of considerable thickness, which have been deposited over relatively short intervals of geological time.

Pore pressure has the effect of reducing any compressive stress acting across a plane, while it has no effect on the magnitude of any shear stress, acting on the same plane. Increasing the pore pressure therefore has much the same result as reducing the confining pressure, favouring brittle fracture rather than ductile deformation at any particular depth.

Classification of Tectonic Structures

Even if the effect of anomalously high pore pressures is taken into account, there is still a transition from brittle fracture to ductile deformation with increasing depth, which occurs in response to changes in the physical conditions at depth. Such a change in the mode of mechanical failure under stress is reflected in the nature of the geological structures which are produced as a result, so providing a basis for classifying these structures according to their morphology.

Extension-fractures are formed by the mechanical failure of rock under the most brittle conditions. They occur as a single set of fractures lying at right angles to a direction of minimum compression in the rock, as shown in Drawing 11. Although tensional stresses are unlikely to be generated at any great

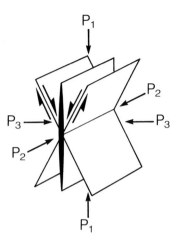

Drawing **11** *Relationship between shear-fractures, lying around 30° to the direction of maximum compression (P_1), and a single extension vein, shown in black, occurring parallel to this direction.*

depth below the earth's surface, owing to the weight of the overlying rocks, it is probably the influence of anomalously high pore pressures which allows extension fractures to form at any depth within the earth. These fractures are mostly preserved in the geological record as *extension-joints*, which have opened up at right angles to their walls. However, mineral veins often occupy extension fractures, so that their formation may well be associated with high fluid pressures, while sheet-like intrusions of igneous rock often follow original planes of weakness in their country-rocks, which presumably opened up under the pressure of the intruding magma.

Shear-fractures mark the start of the transition away from brittle failure towards more ductile behaviour. They are formed by mechanical failure under moderately high confining pressures, provided that the pore pressure remains relatively low. These fractures occur as conjugate sets of shear-planes, making an acute angle with one another of 60° or thereabouts. The line bisecting this angle is parallel to a direction of maximum compression, while the line bisecting the obtuse angle between the two sets of shear-planes is parallel to the corresponding direction of minimum compression. The two sets of shear-planes then intersect one another parallel to a third direction, which lies at right angles to the other two directions of maximum and minimum compression in the rock (see Drawing 11). Unlike extension-fractures, which only split the rock apart without any lateral displacement, shear-fracture involves movements of a shearing nature, in that the rocks on either side of the fracture slip past one another. These movements are directed towards the acute angle between the two sets of shear-planes, away from the obtuse angle. This means that conjugate sets of shear-fractures have the opposite sense of lateral displacement, which can then be used to determine the direction of maximum compression at the time of their formation, as shown in Drawing 12(a). Although such fractures may be preserved in the form of *shear-joints*, they are most obviously represented in the geological record as *faults*. Although breaks are formed as a result of faulting, destroying the original continuity of the rock, there is often effectively no loss of cohesion across the fault-plane.

Shear-zones resemble faults in that the rocks lying on either side of such a structure move past one another. However, these movements are not concentrated on a single plane, or even a number of such planes, cutting through the rock. Instead, they are much more widely distributed throughout the whole

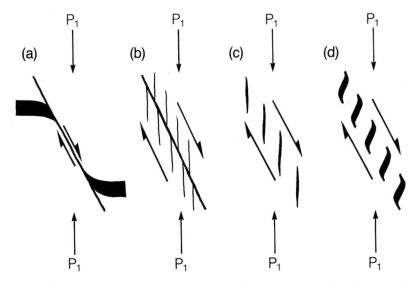

Drawing 12 *How to determine the direction of maximum compression (P_1) from the structural features associated with faults and brittle shear-zones:* (**a**) *displacement on a fault-plane;* (**b**) *feather-fractures along a fault-plane;* (**c**) en échelon *tension gashes;* (**d**) *sigmoidal tension gashes.*

thickness of the shear-zone which is formed as a result. Shear-zones fall into two distinct categories. Firstly, there are *brittle shear-zones* (see **119**), formed by *en échelon* arrays of extension fractures. These often become filled with mineral matter, while the shearing movements may impart a sigmoidal shape to these veins. These two types of brittle shear-zone are shown in Drawings 12(c) and 12(d). Secondly, there are *ductile shear-zones* (see **187**), formed by a relatively narrow zone of intense deformation, flanked on either side by much less deformed rocks, which may indeed show virtually no deformation whatsoever.

Cleavage and *foliation* are typically produced as the result of very much more ductile deformation, affecting the whole mass of the rock. Since this can only occur under the physical conditions appropriate to regional metamorphism, these structures are found in rocks such as slates, phyllites and schists. Their formation involves a wide variety of deformation mechanisms, occurring mostly on a microscopic scale, so that it is only their total effect that can be seen in the field. These effects are most easily observed wherever pre-existing objects of known shape were originally present in the rock as *strain-markers*, recording the nature of the deformation.

Folding and Boudinage

The presence of a lithological layering is not required for reasons of mechanics to allow any of the structural features so far described to form in a rock. However, there are certain other structures which can only affect a rock if such a layering is present. For example, *folds* can only form if there is a pre-existing layering present in the rock, to register the effects of the folding at the very least. However, it is most commonly the presence of lithological differences between separate layers in a rock which allows folds to form in the first place. This occurs as a result of the mechanical instability which develops whenever such a lithological layering is subjected to compression. If the layering is defined by the bedding of sedimentary rocks, which is usually the case, it is possible for the layers to slip over one another on the bedding-planes. This allows folding to occur under relatively brittle conditions. However, a lithological layering typically shows much the same mechanical response to compression under even the most ductile conditions, so that folding is a structural process which reflects a very wide spectrum of mechanical behaviour. It differs in this respect from *boudinage* (see **180**), whereby any relatively brittle layer in a rock breaks up into discrete segments as the result of extension, rather than compression. This can only occur if the surrounding rock behaves in a relatively ductile manner.

Joints and Mineral Veins

Jointing

Nearly every rock is cut by a variety of fractures, along which there has been practically no movement, other than that just needed to open up the fracture in the first place. Such fractures are known as *joints*. It is the lack of any observable displacement which allows joints to be distinguished from faults. Although many joints occur in a non-systematic fashion, appearing to form entirely random patterns, others occur as systematic sets of regularly-spaced fractures, all more or less parallel to one another. Such *joint-sets* are seen to perfection in flat-lying sedimentary rocks, particularly limestones and sandstones. They often combine with one another to form *joint-systems*, as shown in **113**. It is quite common to find that some joints belonging to different sets cut across one another without any deflection, while others end against what are known as *master joints*.

Where two sets of systematic joints are present, they often occur at right angles to one another. This is usually taken to imply that they formed as *extension joints* under an overall state of horizontal tension. The formation of any one joint-set would first relieve this horizontal tension in a particular direction at right angles to itself. Any remaining tension could then be relieved by the formation of a second joint-set, lying at right angles to the first set.

Joints may also occur as two sets, making an acute angle with one another. It has been argued that if such joints are genetically related to one another, they are most likely to represent conjugate sets of *shear-joints*. However, since it is the very nature of joints not to show any evidence of movement, this is often difficult to prove. Circumstantial evidence may be used to suggest such an origin if these joints occur parallel to faults or shear-zones. Otherwise, it seems equally likely that joint patterns of this sort are simply formed by cross-cutting sets of extension fractures, not related in any way to one another.

Surface Markings Some joint-faces display *plumose markings* (Latin: *pluma*, a small feather), formed by a feather-like pattern of low relief, as shown in **114**. The low ridges forming this pattern are thought to mark the directions in which the fracture propagated away from its source. This joint face also shows *conchoidal fractures* (Latin: *concha*, a shell), across which it shows a slight but rather abrupt change in attitude, orientated at right angles to the ridges defining the plumose structure. These fractures therefore trace out a curved path on the surface of the joint. The joint itself may end in a fringe of minor fractures, lying *en échelon* to one another in an oblique angle to the main surface, as shown in **115**. This fringe marks the extreme limits of the joint, beyond which it has not propagated any farther.

Figure 113 *Jointing in a Carboniferous Limestone, forming two sets of vertical joints cutting across one another at right angles, and exposed on a flat-lying bedding-plane. Longhoughton (NU 261159), Northumberland, England.*

Figure 114 *Plumose markings on a joint-surface, forming a series of low ridges which fan out from a central area, so producing a feather-like pattern of low relief. Grantshouse Quarry (NT 811652), Berwickshire, Scotland. (Field of view c 1m)*

Figure 115 En échelon *fractures forming a narrow fringe along a joint-plane. Many joints appear to end in such a fringe of* en échelon *fractures, marking their extremities beyond which they have not propagated any farther. Beacon Hill (SK 510148), Leicestershire, England.*

Extension-veins and Stylolites

Mineral veins fall into two broad categories according to their mode of origin. Firstly, there are *replacement veins* in which the mineral matter has selectively replaced the wall-rocks lying along their course. Such veins are non-dilatational in that the wall-rocks are not offset in any way across the vein. Secondly, there are *dilatational veins*, which are formed by the wall-rocks moving apart to open up a space, which is now occupied by the vein.

Voids may actually form in the rock as a result, which are then filled by the precipitation of the vein materials by mineralizing solutions. They can be recognized by the distinctive presence of *mineral banding*, together with *cockscomb textures*. The latter is formed by the minerals growing at right angles to the walls of the vein, so that they occur as fringes, projecting into its interior. The resulting structure is much the same as *cockade texture* (see **128**).

Some mineral veins occur along faults where step-like changes in the attitude of the fault-plane allow spaces to open up along its course. **116** is a typical example. Other mineral veins are found along joints, providing very useful information about their mode of origin in some cases. For example, **117** shows the presence of needle-like *crystal fibres* in a quartz vein, occupying what is clearly a joint-plane in the field. These crystalline growths are formed by the instantaneous deposition of material along a vein as its wall-rocks move apart, so that there never was any open space to be found along its course. The orientation of these fibres gives the direction in which the walls moved apart. The present case shows that the vein opened roughly at right angles to its trend, giving rise to what can be termed an *extension vein*. Note, however, the small component of sidewards movement also involved in its formation.

Stylolites The effect of what is known as pressure solution on rocks, chiefly limestones, is seen in the development of *stylolites*. **118** shows that they occur in cross-section as highly irregular seams, cutting through the rock in a zig-zag fashion. If the rock is separated into its two parts along such a seam, it is found to consist of a series of striated columns, from which the structure gains its name (Greek: *stylos*, a pillar). Tooth-like projections then interlock with socket-like depressions on the other side, and vice versa.

Stylolites are formed by the highly irregular removal of material along a particular horizon in the rock, leaving a residue of clay minerals and other insoluble matter. This may be demonstrated wherever stylolites cut across fossils or other objects in the rock. They usually follow the bedding, occurring in response to *pressure solution* under the weight of the overlying rocks. However, they are also found at a high angle to the bedding, as the result of horizontal compression.

Figure **116** *Mineral vein occupying a space that opened up along a fault-plane where it changes in dip. Note how the fault-plane can be traced up the cliff-face towards the right, dipping in the opposite direction at a lower angle than the vein itself. Trebarwith Strand (SX 048865), Cornwall, England. (Field of view c 3m)*

Figure **117** *Crystal fibres in a mineral vein, showing how the walls of the original fracture moved apart to allow its infilling with quartz crystals of markedly prismatic a habit. Loose block, Duckpool (SS 201116), Cornwall, England.*

Figure **118** *Stylolites in a Jurassic limestone, forming a highly serrated contact along which material has been removed on either side as a result of pressure solution, acting under the weight of the overlying rocks. Lac de Carcés, Provence, France. (Field of view c 60cm)*

En Échelon *Tension Gashes*

119 shows a typical example of what are known rather loosely as *en échelon tension gashes*, forming an array of mineral veins in the rock. The veins themselves lie parallel to one another at an angle of generally less than 45° to the direction of the array itself. They are termed gashes because they are thickest at their centres, pinching out towards either end in a lenticular fashion. Where crystal fibres define an internal structure to these veins, they are commonly found to cross these veins more or less at right angles. This suggests that the veins open up as extension fractures, so that they are parallel to a direction of maximum compression in the rock.

The staggered arrangement of these extension veins, lying *en échelon* to one another, demonstrates that the array as a whole defines a brittle shear-zone in the rock, lying close to a direction of maximum shear stress. Consider how the rocks on either side of such an array would have moved sidewards in relation to one another as the individual fractures opened up to form the tension gashes. For example, the rocks lying to the left of the arrays shown in **119** would have moved towards the camera, while the rocks lying on the right would have moved away from the camera in the opposite direction. Taken together, these movements can be defined as anticlockwise in their sense of rotation, as it affects the rocks lying within each shear-zone. The direction of maximum compression would then have lain at an acute angle of generally less than 45° to the array itself, as measured in an anticlockwise direction (see **161**).

Arrays of *en échelon* tension gashes may occur as conjugate sets, lying at an oblique angle to one another. They are clearly related to one another if all the tension gashes have an initial orientation in common with each other. It will then be found that the arrays have the opposite sense of rotation in comparison with one another. The direction of maximum compression bisects the acute angle between these conjugate sets of tension gashes, so that it is parallel to the veins forming each set.

120 shows the effect of continued shear on an array of *en échelon* tension gashes. Each vein takes on a sigmoidal shape as its central portion rotates away from its original position, now only preserved by its extremities, close to the margins of the array. If the sense of rotation is clockwise, as shown by the present example, the veins become Z-shaped, while an anticlockwise sense of rotation generates veins that are S-shaped. Often, another generation of tension gashes tends to form, producing a star-like pattern to the individual veins, as shown to a certain extent in **120**. Very occasionally, a reversal in the sense of shear produces tension gashes which lie at right angles to one another, in the manner of **121**.

Figure **119** En échelon *tension gashes forming an array of short and rather stubby veins, staggered in such a way that they lie en* échelon *to one another. These veins are usually filled with quartz or calcite. Baggy Point (SS 435401), Devon, England.*

Figure **120** *Sigmoidal tension gashes lying en* échelon *to one another in a single array, formed by a continuation of the shearing movements which generated this array in the first place. Widemouth Bay (SS 194013), Cornwall, England. (Field of view c 1m)*

Figure **121** *Complex array of* en échelon *tension gashes, formed by a reversal of the sense of movements across the shear-zone as represented by the array itself. Rush Harbour, County Dublin, Ireland.*

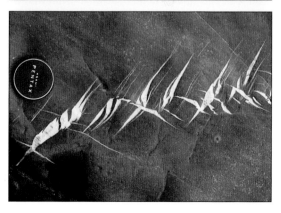

Faults and Faulting

A *fault* is formed wherever movement occurs along a fracture in such a way that the rocks on either side are carried past one another in opposite directions. The total displacement can vary from only a few centimetres, or even less, to much more than a hundred kilometres, depending on circumstances. The break formed as a result is known as a *fault-plane*. Often, the rocks lying on either side of a fault-plane are fractured to a greater or lesser extent, while more than one fault-plane may be present, giving rise to a *fault-zone*.

122 shows a vertical cross-section through a fault, dipping at a moderate angle towards the right. The lateral continuity of the light-coloured beds of sandstone, interbedded with darker shales, has been broken by the fault. What movements occurred on the fault-plane can only be determined where it is possible to match up the beds lying on either side. In fact, the beds identified by the *yellow markers* are one and the same, as can be seen by comparing the sedimentary sequences which occur immediately above and below this horizon on either side of the fault-plane. This means in effect that sedimentary sequences need to be correlated with one another to find out what movements have occurred as the result of faulting. The present instance shows that the *downthrow* was directed towards the right, since the beds lying on this side of the fault-plane are now at a lower level than the equivalent beds on its other side, which have moved up towards the left. It should be emphasized that movements like this on a fault-plane can only be described in relative terms, since it is not known which side actually moved up or down.

The rocks lying in contact with a fault-plane constitute its *wall-rocks*. Provided that the fault is inclined away from the vertical, the rocks lying below a fault-plane form its *foot-wall*, while the rocks occurring above the fault-plane form its *hanging-wall*. For example, the foot-wall occurs to the left of the fault-plane shown in **122**, while its hanging-wall lies to the right. It is the foot-wall which is normally seen wherever a fault-plane is exposed to view, unless it forms an overhang. The typical appearance of such an exposure is shown in **123**.

124 shows another example of a fault, affecting a bedding-plane that dips at a low angle towards the left. Note how the bedding-plane is down-faulted in the same direction by about a metre, forming a somewhat irregular step in its otherwise regular surface. The face of this step forms the foot-wall to the fault. It can be seen that this fault shows quite marked changes in trend along its strike. There is also a subsidiary fault to be seen, downthrowing in the same direction as the main fault. The low cliff lying at the back of the exposure shows the main fault dipping at a high angle to the left, parallel to the staff. How faults can be classified according to the nature of the movements that have occurred on the fault-plane is considered in a later section (see **140** *et seq.*)

Figure 122 *Fault exposed in vertical cross-section, showing only a small amount of downthrow to the right, so that the sedimentary beds lying on either side of the fault-plane can still be correlated with one another. This is a fairly typical example of what is known as a normal fault (see **140**). Howick (NU 259180), Northumberland, England.*

Figure 123 *Fault-plane dipping at a moderately steep angle towards the left, forming a surface that exposes the rocks lying in its foot-wall. The rocks forming the hanging wall to such a fault-plane have been removed as a result of differential erosion. Dunbar (NT 681791), East Lothian, Scotland.*

Figure 124 *Normal fault affecting a bedding-plane, which dips at a low angle towards the left. The fault forms a step-like break in this bedding-plane, across which it is downthrown to the left by about a metre. As the fault-plane dips in the same direction, it represents a normal fault. Clachtoll (NC 036276), Sutherland, Scotland.*

Shatter-zones and Fault-breccias

Although some faults occur as clean-cut fractures, many others are accompanied by disruption and mechanical breakdown of any wall-rocks caught up in the fault-zone. This is marked by shattering, crushing and granulation of the wall-rocks, resulting in the formation of what are known as *fault-rocks*. The mechanical breakdown of pre-existing rocks in response to these processes is generally known as *cataclasis* (Greek: *kata*, completely, and *klastos*, broken).

Shatter-zones are formed wherever rocks become more intensely jointed or fractures as the result of shattering along particular lines of movement. These zones may occur adjacent to fault-planes, or they may be found on their own, forming discrete zones of heavily-jointed rock. As shown in **125**, they often cut across otherwise massive rocks, while they are commonly attacked much more easily by weathering and erosion, so forming hollows in the topography. It is a characteristic feature of shatter-zones that the fracturing of the rocks along their path is not accompanied by any displacement of the fragments formed as a result.

Fault-breccias are a very common type of fault-rock, which results from the fracturing of brittle wall-rocks lying along a fault-zone. This forms a jumbled-up mass of angular fragments, often cemented together by mineral matter. **126** shows a typical example, formed by the fracturing of a well-cemented sandstone adjacent to a small fault. Note how the individual fragments can be pieced together, so allowing their original outlines to be determined, even though they are now separated from one another by brown-weathering dolomitic cement. There must have been an increase in volume resulting in the dilatation of the rocks affected by the faulting, perhaps aided by high fluid pressures exerted by the mineralizing solutions in flowing through the fault-zone. **127** is another example of a fault-breccia, showing a much greater degree of fracturing.

Figure 125 *Shatter-zone cutting the igneous rocks of the Whin Sill in the North of England. The shattered rocks are attacked more easily by weathering and erosion than the less fractured rocks in the background, so that they form a hollow in the topography. High Force (NY 880284), County Durham, England.*

Figure 126 *Fault-breccia formed by angular fragments of sandstone, cemented together in a higgledy–piggledy fashion by brown-weathering dolomite. It has been formed by the fracturing of a quartzitic sandstone adjacent to a small fault. Salisbury Craigs (NT 266732), Edinburgh, Scotland.*

Figure 127 *Fault-breccia consisting of finely-comminuted fragments of Lewisian Gneiss, lying in a matrix formed by even more finely-fragmented material. The wall-rocks on either side of this fault-breccia are formed by Lewisian Gneiss, unaffected by the movements. Achmelvich Bay (NC 053260), Sutherland, Scotland.*

The fragments in a fault-breccia are frequently cemented together by mineral matter, consisting of quartz, calcite, dolomite or barytes in most cases, which lacks any well-developed structure. Occasionally, however, as shown in **128**, this mineral matter occurs in fringes of well-formed crystals projecting into the spaces around each fragment, giving rise to *cockade texture*. This can only happen if voids open up between the fragments in a fault-breccia. Crystals are then precipitated from the mineralizing solutions as crusts around each fragment, creating fringes of crystal faces which project into the spaces between the fragments.

Although breccias often consist of angular fragments cemented together by mineral matter, these fragments can also be so affected by crushing and more extreme granulation that any remaining fragments may be set in a matrix of finely comminuted material. **129** shows what can be taken as the initial stages in such a process, where a granitic rock has been fractured on such a fine scale that its original texture has largely been destroyed. *Microbreccias* are formed as a result wherever cataclasis has reduced the grain-size of the original rock to less than a millimetre. Such a rock still retains a hard and compact character. It differs in this respect from *fault-gouge*, which is a rock of clay-like consistency commonly found along fault-planes. It has a gritty texture when dry, becoming soft and sticky when wet. It is formed by the effects of crushing on relatively weak rocks, such as shales.

Fault-planes are often lined with discrete zones of fault-breccia, making sharp contacts with the wall-rocks on either side. Other examples may, however, show more transitional contacts, with only a gradual reduction in the amount of brecciation affecting the wall-rocks as they are traced away from the fault-zone.

Crush-breccias occur wherever the rocks lying within a fault-zone are affected by crushing and shearing, rather than fracturing of a more brittle nature. This produces rounded or lenticular fragments, which are surrounded by rather more crushed and sheared material. Although such a rock might best be termed a crush-conglomerate, the use of the latter term to describe a rock not of sedimentary origin is thought to be inappropriate. **130** shows an example of a crush-breccia, which is associated with a fault-zone that trends horizontally across the field of view. The steeply-dipping rocks occupying the foreground are cut off by the crush-breccia along the margin of the fault-zone itself. Note how the crushing and shearing which affects the rocks lying within the fault-zone has produced a crude layering parallel to its margins.

Pseudotachylyte Veins and Breccias

The frictional heating that can occur as the result of extremely rapid fault-movements may be sufficient to cause the rocks lying within a fault-zone to melt. The molten material may then intrude the wall-rocks as irregular veins and stringers of a rock known as *pseudotachylyte*. This is a dark and extremely fine-grained rock resembling tachylyte (Greek: *tachys*, swift, and *lytos*, melted or fused), which is a type of basaltic glass. It gains its name from the fact that it is easily fused in the laboratory. Pseudotachylyte differs from *flinty crush-rock* in its intrusive habit, together with the absence of any structures produced by local cataclasis. In other words, flinty crush-rock is more commonly found along major planes of movement, while pseudotachylyte intrudes the wall-rocks away from such planes of movement. Typically, pseudotachylyte carries fragments of finely comminuted fault-rocks, while it occupies ramifying networks of irregular fractures which may intersect one another to isolate angular blocks of the wall-rocks, so forming *pseudotachylyte breccia*.

Figure 128 *Cockade texture in a fault-breccia, formed by the growth of prismatic minerals into the spaces formed between the fragments of wall-rock lying within a fault-zone. Strontian (NM 823659), Inverness-shire, Scotland.*

Figure 129 *Cataclastic texture in the Strontian Granite adjacent to the Great Glen Fault in the North of Scotland. Note how the original texture of this coarsely crystalline rock has been almost entirely destroyed as a result. Loch Linnhe (NM 885540), Inverness-shire, Scotland.*

Figure 130 *Crush-breccia formed by crushing and shearing of country-rock fragments within a fault-zone. This typically results in rounded or lenticular fragments, surrounded by rather more crushed or sheared material. Laxo Voe (HU 459636), Shetland, Scotland. (Field of view c 2.5m)*

Mylonites and Flaser-gneisses

Mylonites (Greek: *mylon*, a mill) are compact and very fine-grained rocks, often with a cherty appearance, produced as the result of extreme deformation within major fault-zones. They typically show a platy lamination or foliation, otherwise known as *fluxion-structure*, which is often accompanied by a fine compositional banding. As shown in **131**, this banding is often folded on a small scale. The presence of fluxion-structure serves to distinguish mylonites from microbreccias, which lack such a directional fabric.

When mylonites were first described, it was thought that they were formed by extreme comminution of the wall-rocks affected by the faulting. According to this view, as reflected in the name originally given to these rocks, they consist of broken-down fragments of the mineral grains initially present in the rock, reduced to a rock-flour by the effects of mechanical grinding between two surfaces, as in a corn-mill. However, it now appears that, although mineral grains such as the feldspars may be so affected, the greater proportion become drawn-out as the result of plastic deformation, rather than brittle fracture.

In particular, fluxion-structure is the product of plastic deformation affecting the individual grains, which become permanently distorted as the result of various mechanisms occurring on an atomic scale. The extremely fine-grained nature of the rock is then simply a result of incipient recrystallization, which has replaced the original grains in the rock, now highly strained as a result, with myriads of very much smaller grains, all showing the effects of much less strain. If this process is repeated time and time again during the course of deformation, mylonites eventually merge into *ultra-mylonites*, so deformed and recrystallized that fluxion-structure can hardly be recognized, at least in hand-specimen.

Flaser-gneisses are formed wherever the various minerals in a coarse-grained rock respond in different ways to the processes of deformation and recrystallization, which are otherwise involved in the formation of mylonites. Commonly, it is the feldspars which offer the most effective resistance to these processes, in contrast to quartz and the micas, which are more easily deformed. **132** shows a sheared granite in which the feldspar grains occur as remnants, lighter in colour than the more deformed matrix surrounding them. It is the more easily-deformed minerals which are reduced to trails of more finely-grained material, winding between lenticular masses of the original rock, so forming a flaser-gneiss. Further deformation tends to convert such rocks into *augen-mylonites* (German: *augen*, eyes), as shown in **133**. Note how the feldspar grains occur in the form of lenticular *augen*, lying within a mylonitic foliation, which forms the structure of the surrounding rock.

Figure **131** *Mylonite showing the typical appearance of this extremely fine-grained cataclastic rock, formed by intense deformation affecting the rocks lying within a fault-zone. Mylonites have a finely laminated, banded or streaky texture, often showing the effects of flow-folding, which is a characteristic feature of such rocks. Ullapool (NH 160915), Wester Ross, Scotland.*

Figure **132** *Flaser-gneiss formed by a sheared granite, in which lenticular remnants of the original feldspar grains are separated from one another by a much finer-grained groundmass, itself foliated and consisting mostly of quartz and mica. Barnesfjorden, Jotunheim, Norway.*

Figure **133** *Augen-mylonite formed by even more extreme deformation than affected the flaser-gneiss shown in the previous photograph, resulting in lenticular augen of feldspar, set in a mylonitic groundmass. Playa de Xilloy, Lugo Province, Spain. (Photograph by C.T. Scrutton)*

Slickensides and Related Structures

The surface of a fault-plane is often rendered smooth and polished by the frictional movements. **134** shows a typical example. *Slickenside* is the term applied to such a surface, derived from the Middle English (*slicken*, to render smooth). However, this surface is commonly striated in the direction of movement on the fault-plane, so that the striations themselves are sometimes known erroneously as slickensides, under the mistaken impression that the term itself implies the notion of sliding. These striations are simply scratches on the polished surface, gouged out by any irregularities on the opposite surface of the fault-plane. Commonly, they are formed by angular fragments of the wall-rocks, embedded in more broken-down material lying along the fault-plane.

Although they have the same trend, such striations should be distinguished from the *lineated fault-veins* that can occur wherever faulting is accompanied by mineralization along the fault-plane. **135** shows a typical example, formed by the growth of prismatic minerals such as quartz, calcite or serpentine, so orientated that their long axes are all arranged parallel to one another. This fabric can only be formed where potential voids open up along the fault-plane as a result of the two walls moving slightly apart during the course of the faulting. This may be caused by the presence of slight steps or other irregularities along the course of the fault-plane, or it may be a reflection of movements occurring at a slight angle to the trend of the fault-plane itself. The prismatic minerals often appear to nucleate on minerals of the same species wherever they form the wall-rocks immediately adjacent to the fault-plane. Each mineral fibre would then join up the points on the opposites of the fault-plane that were originally in contact with one another.

135 also shows that the surface of such a lineated fault-vein is often exposed as a series of small steps, cutting across the mineral fibres at right angles. Given the internal structure of these veins, these steps will all face the same direction, which then corresponds to the direction of movement as shown by the opposite wall of the fault-plane. Likewise, although experimental evidence has cast some doubt on the matter, it is often thought that a slickenside surface feels smoother to the touch when it is rubbed in the direction of movement, as shown by the side now removed by erosion to expose the fault-plane.

Since continued movements along a fault will tend to destroy any structures already present along its course, striated and lineated surfaces within a fault-zone can often only be taken as evidence concerning the nature of the very latest movements. Occasionally, however, as shown in **136**, there is evidence for a gradual change in the direction of movement.

Figure **134** *Striations on the slickenside surface of a fault-plane, which has been rendered smooth and polished as a result of the movements. These striations give the direction in which the wall-rocks on either side of the fault-plane moved with respect to one another as a result of the faulting. Kalamafka, Crete.*

Figure **135** *Mineral fibres formed in a fault-vein system where the growth of prismatic minerals has occurred at the same time as the walls of a fault-plane moved apart during the course of the faulting. Grantshouse Quarry (NT 811652), Berwickshire, Scotland.*

Figure **136** *Curved striations on a fault-plane, formed where the direction of movement has changed during the course of the faulting. Mineral fibres may also show a curved form under these circumstances, or more than one set may be present, differing from one another in orientation. Colmenar, Malaga Province, Spain.*

Fault-drag and Rollovers

The nature of the movements that occurred on a fault-plane can be inferred wherever *fault-drag* has affected the wall-rocks. It is usually the bedding of sedimentary rocks which registers the effects of faulting in this way, as shown in **137**. The fault-plane in the present example dips towards the left at a moderately steep angle, while the bedding dips much more steeply towards the right. Note how the bedding is deflected, curving towards the line of the fault-plane as a result of the movements. Clearly, it is the rocks forming the hanging-wall on the left, which have moved upwards in comparison with the rocks forming the foot-wall on the right.

The opposite effect is illustrated in **138**, which shows a structure known as a *listric fault* (Greek: *listron*, a shovel). Such a fault has a curved trace, at least in cross-section. This is seen in the present example, which changes in attitude along its course in such a way that it eventually comes to lie along the bedding of the sandstone.

Unless it forms a circular arc, any movements taking place along such a curved fault-plane tend to open up a gap between its walls. However, a potential void obviously did not form in the present instance, simply because the wall-rocks became distorted as a result of the movements. This distortion produces a structure known as a *rollover* in flat-lying sediments, marked by

the bedding coming to dip towards the fault-plane on its downthrown side. *Reverse fault-drag* then occurs as the result of the hanging wall collapsing against the foot-wall.

Feather Fractures

The close association between extension fractures and shear-zones, which is seen in the development of *en échelon* tension gashes, is also found in the formation of what are known as *feather fractures*. While they may form as extension joints, parallel to a direction of maximum compression in the rock, feather fractures are often filled with vein minerals. This is shown in **139** by the set of slightly brown-weathering veins, which occur as feather fractures forming a fringe to the other vein, identified by its lighter colour and greater width. This vein represents a fault now filled with vein material. Mechanical considerations indicate that the acute angle between the feather fractures and the fault-plane points in the direction of movement on the fault-plane itself, as shown in Drawing 12(b). This means that the rocks forming the upper part of the field of view have moved towards the right, in comparison with the rocks lying on the other side of the fault-plane, which have moved towards the left.

Figure **137** *Fault drag deflecting the bedding of wall-rocks immediately adjacent to a fault-plane. As the faulting has evidently caused the overlying rocks to move up the dip of the fault-plane, the fault itself represents what is known as a reverse fault (see* **143**)*. Kyrenia, Cyprus.*

Figure **138** *Listric fault affecting a bed of sandstone, immediately to the right of the coin. The fault-plane forms a curved surface, which changes in dip in such a way that it eventually merges with the base of this sandstone bed. Broad Haven (SM 860142), Dyfed, Wales.*

Figure **139** *Feather fractures forming a set of slightly brown-weathering veins, lying at an acute angle to another vein, which trends almost horizontally across the field of view. The fault-plane represented by this vein was formed in response to a maximum compression, trending parallel to the direction of the feather-fractures. Cracklington Haven (SX 14968), Cornwall, England.*

147

Dip-slip and Strike-slip Faults

While the attitude of a fault-plane may be determined easily enough in the field, provided that it is sufficiently well-exposed, it is often much more difficult to decide how the rocks affected by the faulting have moved in relation to one another, without making certain assumptions. Nevertheless, it is commonly assumed that most faults can be divided into two broad categories. *Dip-slip faults* include all those faults where the movements have occurred up or down the fault-plane, parallel to its dip. Such faults can then be distinguished from *strike-slip faults*, which involve horizontal rather than vertical movements, occurring parallel to the strike of the fault-plane.

Normal Faults

As the name suggests, *normal faults* provide the most common examples of dip-slip faults. They can be recognized wherever it is the hanging-wall that has been downthrown, relative to the foot-wall. This means in effect that such a fault can only be recognized if the direction of its dip can be determined. **140** then shows how normal faults typically occur at a high angle, dipping at 60–70° from the horizontal. Three fault-planes can be identified, each dipping to the right, while a conjugate fault is also present, farther to the right, dipping somewhat more steeply in the opposite direction. Note that the faulting has resulted in horizontal extension within the plane of the exposure.

It is the presence of conjugate sets of normal faults, dipping in opposite directions to one another, which results in *block faulting*. Downthrown rocks lying between two normal faults, which dip in towards one another, then define a structure known as a *graben* (German: *graben*, a ditch), particularly if developed on a large scale. **141** shows such a down-faulted block, lying between two normal faults. A

horst (German: *horst*, an eagle's nest) is the exact opposite to a graben, formed by an upthrown block of rocks lying between two normal faults, which dip away from one another in opposite directions.

Step faulting occurs wherever only a single set of normal faults is present, all dipping in the same direction. **142** shows a series of very closely-spaced step-faults, cutting across a quartzite. Although they appear at first sight to be joints, examination of the somewhat darker rock forming a band towards the bottom of the exposure shows that these fractures are normal faults, all downthrowing in the same direction towards the left.

Reverse Faults

Any dip-slip fault showing the opposite sense of displacement to a normal fault is known for obvious reasons as a *reverse fault*. As shown in **143**, it is then the hanging wall that has moved upwards rather than downwards, relative to the foot-wall. Although it may be said that the light-coloured beds of sandstone have been thrust over one another from the left, it is equally possible that underthrusting has occurred from the right. Whatever mechanism is responsible for the faulting, it is still the case that the beds have undergone a certain degree of shortening within the plane of the exposure as the result of horizontal compression. This is reflected in its turn by the duplication which is shown by the sandstone beds where they overlap one another above and below the fault-plane. All these relationships are the reverse of that shown by normal faults.

Reverse faults are generally known as *overthrusts*, wherever they occur on a large scale as low-angle structures, close to the horizontal. They often make up the most important features of major thrust-zones, as described in the next section. However, other types of thrust-fault are found, often

Figure 140 *Normal faults affecting a sequence of flat-lying sandstones, showing how the downthrow on such faults occurs in the same direction as the dip of the fault-planes. This is best seen where the faults cut three prominent beds of sandstone, exposed towards the top of the cliff-section. Widemouth Bay (SS 197038), Cornwall, England. (Field of view c 8m).*

Figure 141 *Normal faults as seen in close-up from the same locality as shown in the previous photograph, showing their conjugate nature. A downthrown block is present, forming a miniature graben, flanked by two normal faults which in dipping towards one another each show the opposite sense of displacement. Widemouth Bay (SS 197038), Cornwall, England.*

Figure 142 *Step-faults forming a series of very closely-spaced fractures, all parallel to one another, and downthrowing consistently towards the left, as shown by their effect on the band of darker rock towards the bottom of the exposure. Onich (NN 042611), Inverness-shire, Scotland.*

closely associated with folding. An example of such a thrust-fault is shown in **144**. The fault-plane in question dips at only a few degrees to the right, while the hanging-wall is displaced in the opposite direction through a distance of not much more than a metre, as judged by its effect on the light-coloured beds of sandstone. Such a fault is known as a *break-thrust*, assuming that it forms once the folding cannot accommodate any further degree of horizontal compression.

High-angle reverse faults lie at the opposite extreme to overthrusts, forming what are known as *upthrusts*. They typically have fault-planes close to the vertical, while they are often found to affect the steep limbs of monoclinal folds. The fault shown in **6** is a good example of an upthrust. The closer view of this structure provided by **145** shows that the fault-plane is itself exposed as the scarp of light-coloured rock, strongly affected by the topography, which dips at a steep angle *towards* the left. Its hanging-wall is therefore formed by the metamorphic rocks occupying the foreground. Since these rocks must be older than the flat-lying limestones of Mesozoic age in the background, forming the foot-wall, it can be assumed that the hanging wall has moved up the dip of the fault-plane, relative to the foot-wall. If this is the case, the fault is an upthrust, bringing up the metamorphic rocks from greater depths, where they originally underlay the Mesozoic limestones. It is perhaps unusual in that the bedding of the Mesozoic limestones has not been affected in any way by the faulting.

Strike-slip Faults

The only other category of any importance, apart from dip-slip faults, is represented by strike-slip faults, which show the effects of horizontal rather than vertical movements. Ideally, strike-slip faults occur as conjugate sets of vertical fractures, trending at an angle of 60° or thereabouts to one another. **146** shows a vertical bed of rather massive sandstone, forming a particularly resistant horizon, which is cut by several strike-slip faults.

The movements on strike-slip faults can be described as sinistral (left-lateral) or dextral (right-lateral), according to whether the rocks on the far side of the fault-plane have moved left or right. This terminology can be applied from whatever side the fault-plane is viewed, since the far side always appears to have moved in the same direction. **146** therefore shows a series of sinistral or left-lateral strike-slip faults.

Figure **143** *Reverse fault affecting a series of sandstone beds in such a way that the rocks forming its hanging wall have apparently been thrust up the fault-plane from the left in the opposite direction to the dip of the fault-plane itself. Howick (NU 259180), Northumberland, England.*

Figure **144** *Break-thrust cutting across the vertical limb, nearly but not quite overturned, which lies between the anticline and its underlying syncline. The thrusting was directed towards the left in the same direction as the folding. Broad Haven (SM 860142), Dyfed, Wales. (Height of section c 5m)*

Figure **145** *Upthrust between metamorphic rocks forming the rather featureless foreground, and the well-bedded Mesozoic limestones which form the summits of the hills. Since the fault-plane dips steeply towards the left, it is the Mesozoic limestones which make up the footwall to this fault, so identifying the structure as an upthrust. Lastros, Crete.*

Figure **146** *Strike-slip faults affecting a vertical bed of rather massive sandstone so that it now forms a series of isolated fault-blocks. The faults themselves trend across the field of view almost at right angles to the strike of the bedding. Stonehaven (NO 884872), Kincardinshire, Scotland. (Field of view c 10m)*

THRUST-ZONES AND DUPLEXES

The margins of fold mountains are often defined by major zones of overthrusting. The movements are typically directed outwards from the central parts of the orogenic belt, towards what is termed its *foreland*. Such overthrusting usually affects thick sequences of sedimentary rocks, which accumulated rapidly on top of an older basement forming the foundations to the orogenic belt in the early stages of its evolution. The thrust-planes typically have a "staircase" geometry, consisting of a series of *bedding-plane thrusts*, which are connected with one another by *ramps*, where the thrust-planes cut up across the stratigraphy at angles of around 30°, as shown in Drawing 13. The *flats*, as these bedding-plane thrusts are otherwise known, tend to follow mechanically weak horizons, formed in particular by evaporites, shales or dolomites, wherever they are interbedded with more rigid rocks, such as sandstones and limestones. The overlying rocks then become detached by thrusting along these mechanically weak horizons, so forming what are known as surfaces of *décollement* (French: *décoller*, to unglue). The presence of bedding-plane thrusts, separated from one another by intervening ramps, therefore suggests that several horizons of potential décollement are present within the stratigraphic sequence as a whole.

Since overthrusts typically ascend the stratigraphic sequence in a series of structural ramps as they are traced towards the foreland, they nearly always bring older rocks to rest on top of younger rocks, contrary to the Principle of Superposition. The older rocks resting on top of such an overthrust constitute what is known as a *thrust-sheet* or a *nappe* (French: *nappe*, a table-cloth). If they are sufficiently far-travelled, so that they are now out of place in comparison with the underlying rocks, they can be described as *allochthonous* (Greek: *allos*, another, and *chthōn*, country). The rocks belonging to the foreland, which have not been transported from their place of origin, may then be described as

Drawing **13** *An overthrust formed by bedding-plane "flats", separated from one another by a ramp in the thrust-plane. Note how the thrust-plane climbs to higher stratigraphic levels in the direction of overthrusting, placing older rocks on top of younger ones as it does so.*

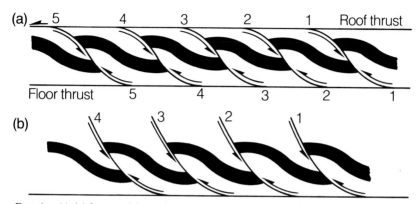

Drawing 14 (a) *Structural duplex formed by a series of minor thrusts sandwiched between a floor thrust and a roof thrust.* (b) *Imbricate stack of minor thrusts lying above a major thrust-plane.*

autochthonous. Any outlier of allochthonous rocks, resting on a thrust-plane so that it is entirely surrounded by the underlying rocks, forms what is known as a *klippe* (German: *klippe*, a crag, plural, *klippen*), detached from its main outcrop. If these relationships are reversed, so that the underlying rocks are exposed to view, completely surrounded by allochthonous rocks, a *tectonic window* would be present in the otherwise continuous outcrop of the overlying thrust-sheet.

Overthrusts often combine with one another to form what are now known as *duplexes*, formed wherever two thrust-planes occur at different structural levels, as shown in Drawing 14(a). This means that each duplex consists of a *roof thrust* and a *floor thrust*, between which a whole series of minor thrusts often occur in an imbricate fashion (Latin: *imbrex*, a roof-tile). Each minor thrust-plane curves upwards from the floor thrust into a moderately steep attitude, before flattening out again as it merges with the roof thrust. If a roof thrust is absent, such a splay of minor thrusts, curving upwards from a major thrust-plane, is known as an *imbricate stack*, as shown in Drawing 14(b). Several duplexes may be present within a major zone of overthrusting, with the basal thrust forming what is termed the *sole thrust*.

The sequence of movements within a major thrust-zone is a matter of some controversy. Traditionally, it was thought that the uppermost thrusts were the last to form, riding up over the backs of the earlier thrusts once the latter encountered any obstacles to their further movement. However, it can equally well be argued that the earlier thrusts have a snow-plough effect, causing the latter thrusts to propagate at lower levels towards the foreland through the progressive collapse of foot-wall ramps. This model implies that it is the lowermost and most forward thrusts which were the last to form, as the thrusting extended outwards in the direction of the foreland.

Major Overthrusts in the Field

The Moine Thrust-Zone can be taken as an example of a major zone of overthrusting, although it is slightly unusual in that basement rocks are affected by the movements, as well as their sedimentary cover. It forms the northwestern margin of the Caledonian Fold-Belt in the NW Highlands of Scotland. Several overthrusts are present within this thrust-zone, which dips at a low angle towards the ESE, away from the foreland to the Caledonian Fold-Belt, which lies farther to the northwest. The basement rocks belonging to this foreland are represented by the Lewisian Gneiss, dating back to the Early Precambrian, overlain unconformably by the Torridonian Sandstone, belonging to the Late Precambrian. Cambro-Ordovician rocks rest unconformably on top of these basement rocks, so forming a sedimentary cover to this basement. All these rocks are found as thrust-sheets within the Moine Thrust-Zone, underlying the Moine Thrust itself.

The Glencoul Thrust provides a well-exposed example of such a thrust-sheet, which is shown in **147**. The lowermost slopes of Beinn Aird na Loch, the prominent hill on the sky-line, are occupied by Lewisian Gneiss. The basement rocks are overlain unconformably by the Cambro-Ordovician sediments, forming a pronounced feature which runs downhill towards the right. The sole thrust lies just above this feature, so that all the underlying rocks belong to the NW foreland of the Caledonian fold-belt. A thin zone of imbrication, although not always present,

occurs immediately above the sole thrust, overlain in its turn by the Lewisian Gneiss of the Glencoul Thrust. This forms the rugged but rather featureless ground towards the summit of Beinn Aird na Loch, similar to the exposures found farther downhill to the left. The Glencoul Thrust has therefore brought Lewisian Gneiss of the basement to rest on top of Cambro-Ordovician sediments of a much later date. The overthrusting was directed from the right towards the foreland.

148 shows the Glencoul Thrust, where it is exposed on the other side of the loch, so that it appears to dip in the opposite direction. Note that there is a very sharp contact between the Cambro-Ordovician rocks, forming the light-coloured dolomites below the thrust-plane, and the much darker rocks on top, representing the Lewisian Gneiss of the Glencoul Thrust-Sheet, now affected by a certain degree of mylonitization.

The Moine Thrust is itself shown in **149**. Again, a clean-cut thrust-plane is underlain by dolomites of Cambro-Ordovician age. They are overlain by metamorphic rocks of the Caledonian Fold-Belt, now somewhat mylonitized, thrust into their present position from the ESE to form the Moine Nappe. Elsewhere, thick bands of mylonitic rocks are found immediately below the Moine Thrust, overridden by these metamorphic rocks (see **131**). These mylonites commonly carry a stretching lineation (see **197**), which indicates that the thrusting was directed towards the WNW.

Figure **147** *Glencoul Thrust forming part of the Moine Thrust-Zone in the north of Scotland. The Lewisian Gneiss forming the rough ground below the prominent feature which can be traced down the hillside to the right is repeated by the thrusting so that it is also exposed above this feature, marking the line of the Glencoul Thrust. Unapool (NC 236318), Sutherland, Scotland.*

Figure **148** *Exposure of the Glencoul Thrust forming a tectonic contact between the light-coloured dolomites of Cambro-Ordovician age, which lie underneath the thrust-plane and the much older rocks lying on top of the thrust-plane, formed by the Lewisian Gneiss at the base of the Glencoul Thrust-Sheet itself. Loch Glencoul (NC 259302), Sutherland, Scotland. (Height of section c 5m)*

Figure **149** *Exposure of the Moine Thrust at Knockan Craig in the North of Scotland, marking a sharp contact between mylonitic rocks of the overlying Moine Nappe and the underlying dolomites and limestones of Cambro-Ordovician age. Knockan Craig (NC 190091), Wester Ross, Scotland. (Height of section c 5m)*

155

FOLDS AND FOLDING

Vertical Movements Some folding occurs wherever the vertical movements affecting flat-lying sedimentary rocks vary in amount. This may simply be a consequence of differential compaction. This tends to occur wherever there are lateral variations in the degree of compaction that can be undergone by sedimentary rocks which were originally deposited side by side with one another. *Compaction folds* are formed as a result, often occurring over buried hills in an underlying basement, as shown in **105**. However, it is equally possible for flat-lying rocks to be affected by vertical movements as the result of tectonic forces. This commonly produces rather irregular structures in the form of domes and basins, lacking any particular trend. They are formed by warping and tilting of sedimentary strata which now dip in a variety of different directions. *Domes* are formed by sedimentary strata dipping away in all directions from a central area, where the oldest rocks are now exposed as the result of erosion, surrounded by progressively younger rocks in a series of roughly concentric outcrops. *Basins* are the exact opposite, formed by sedimentary rocks dipping in towards a central area from all directions. The youngest beds are then exposed in the centre of the basin, surrounded by progressively older rocks which again form a series of roughly concentric outcrops around this central area. Domes and basins exhibiting a more elongate form are sometimes known as *periclinal folds*. All these structures tend to occur as broad and rather irregular features, many kilometres across, without any common trend.

The form shown by domes and basins may be accentuated by the effects of faulting. For example, the vertical movements accompanying the dip-slip faulting of basement rocks often extend upwards so that folding rather than faulting affects an overlying cover of sedimentary rocks. Even within such a sedimentary cover, folding may take the place of faulting as a partial expression of vertical movements, which become more widely distributed as a result. Any changes in downthrow along the strike of particular faults, showing dip-slip movements, must also cause folding to affect the intervening rocks. Such folding will vary in its trend according to the pattern of faulting, particularly as it affects an underlying basement. Moreover, if evaporites are present as thick deposits of rock salt, particularly at lower levels within the sedimentary sequence, the upward movement of this highly mobile rock of relatively low density under the influence of gravity will generate further structures, known as *salt-domes*, often adding a further complication to the structural pattern developed in response to vertical movements. Similar effects can be exerted by the forceful intrusion of igneous rocks, resulting in the folding of their country-rocks.

Lateral Compression Apart from such structures, formed in response to vertical movements, the folding of sedimentary rocks is mostly a consequence of horizontal compression, acting parallel to the bedding. If the physical conditions favour relatively ductile deformation, this may simply cause the layers to undergo a shortening along their length, counterbalanced by a gradual increase in their thickness. This process is known as *layer-parallel shortening*. It leads eventually to the development of a cleavage within the layer so affected, once the deformation becomes sufficiently large. Folding only occurs if a horizontal compression leads to mechanical instability between the layers, causing them to buckle into a wave-like form. Such buckles will only form if there are differences in lithology between the layers, however subtle, influencing their response to lateral compression. These differences in mechanical behaviour are reflected in what is known as the *competency* of the layers. It is the competent layers which are relatively more difficult to deform than the incompetent layers, so that they control the form of the folds which develop as a result of buckling or any other mechanism. Typically, sandstones, limestones and dolomites form the more competent layers within a sedimentary sequence, while shales and evaporites form the relatively incompetent layers.

Although buckling is a mechanism which can cause folding to occur under a wide variety of physical conditions, it is possible to recognize two separate zones within the earth's crust where folding takes on a reasonably distinctive character. Firstly, sedimentary sequences may be folded at relatively shallow depths, where the physical conditions of temperature and confining pressure do not allow any metamorphic reactions to occur. This means that very little if any layer-parallel shortening takes place before the individual folds start to develop in response to lateral compression. Such folding affects even relatively brittle rocks simply because the layers are able to slip over one another along the bedding-planes which separate the folded layers from one another. The folds produced as a result commonly exhibit a structural style of relative simplicity, unless it is complicated by the effects of faulting. Secondly, folding can also take place along more deep-seated zones within the earth's crust, corresponding to the roots of orogenic belts, where the temperatures and confining presures are sufficiently high for metamorphism to occur on a regional scale. These conditions allow a considerable degree of layer-parallel shortening to occur before the individual folds start to form as a response to the buckling of the more competent layers under lateral compression. This means that cleavages are often developed, closely associated with the folding of relatively ductile rocks under conditions which allow a considerable degree of deformation to occur. The resulting structures may be exceedingly complex, particularly where more than one phase of deformation and regional metamorphism can be recognized, affecting rocks which were first folded and then refolded in response to a complex history of orogenic movements.

Geometry of Cylindroidal Folds

It is possible to describe the geometrical form of a folded surface in a relatively simple manner if its cross-section remains much the same along its entire length, so that it makes up what is known as a *cylindroidal fold*. This then consists of a pair of *fold-limbs*, separated from one another by what is termed the *fold-hinge*. A typical example is shown in **150**. The fold-hinge lies parallel to the walking stick, separating the fold-limbs from one another. The bedding has much the same dip and strike throughout each fold-limb, while it is folded around the fold-hinge, where it changes more or less abruptly in its attitude. The *hinge-line* is then defined as the line lying parallel to the fold-hinge, across which the folded surface shows its greatest degree of curvature in passing from one fold-limb to the other. If there is a rather gradual change in attitude across the fold-hinge, the fold may then be said to possess a *hinge-zone*, separating the fold-limbs from one another, as shown in the present instance.

The direction of least curvature within a folded surface is parallel to what is known as the *fold-axis*. It may be taken as the closest approximation to a straight line which, if moved parallel to itself, would generate the folded surface. This could be done in **150** by moving the walking stick in such a way that it sweeps out the form of the folded surface while remaining parallel to itself. Defined in this manner, the fold-axis has no position in space, since it is merely a direction. It differs in this respect from any fold-hinges lying within such a folded surface, which have a certain position as well as defining a particular direction, parallel to the fold-axis.

The true shape of a cylindroidal fold can only be seen if the folded layers are viewed in the same direction as the fold-axis, parallel to its *plunge* (see page 166). Although this provides a cross-section within a plane lying at right angles to the fold-hinge, it is not necessary for the folded layers to outcrop within such a plane. For example, **151** shows the outcrop of a series of folded layers, exposed on a steep hillside. By descending into the valley, this fold can be viewed up its plunge, as shown in **152**, to reveal its true cross-section. Normally, of course, such a cross-section can only be viewed down-plunge, providing a foreshortened view of what is known as the *fold-profile*.

Figure 150 *Cylindroidal fold as seen in an oblique view, showing how it consists of a fold-hinge parallel to the shaft of the walking stick, together with a pair of fold-limbs lying on either side of this fold-hinge. This fold plunges at a low angle towards the right. Invermoriston (NH 419169), Inverness-shire, Scotland.*

Figure 151 *Outcrop pattern formed by a fold plunging slightly less steeply than the slope of the hillside on which it is exposed. This provides a distorted view of the fold-profile since its outcrop is affected by the topography. Elburz Mountains, Iran.*

Figure 152 *Profile of the fold shown in the previous photograph, seen by descending into the valley and then looking up its plunge. This represents the true shape of the folded surfaces, as viewed in cross-section within a plane at right angles to the plunge of the fold-axis. Elburz Mountains, Iran.*

Folds and their Description

Plunge The orientation of a cylindroidal fold can first be described in terms of its *plunge*. This refers to the angle made by the fold-hinge, as measured downwards from the horizontal within a vertical plane, using a clinometer. The trend of the fold is then given by the direction of its plunge, measured as a bearing from true north. Finding the plunge allows the following categories to be recognized: *horizontal folds* plunging at less than 10° from the horizontal; *plunging folds* with fold-hinges plunging at angles between 10° and 80°; and *vertical folds*, plunging at more than 80° from the horizontal.

Inclination Just finding the plunge is not enough to describe fully the orientation of a cylindroidal fold, since another direction is also needed. This is given by the *axial plane*, defined simply as the surface which separates the fold-limbs from one another. The axial plane therefore passes through each fold-hinge in its turn, as it is traced through a fold. The *inclination* of a fold then refers to the dip of its axial plane. *Upright folds* have axial planes dipping at more than 80°. They are sometimes known as symmetrical folds, since they may have fold-limbs dipping at much the same angle in opposite directions to one another (see **153**). *Inclined folds* have axial planes dipping at angles between 80° and 10°. They are sometimes known as asymmetrical folds (see **154**).

Overfolds are formed wherever one fold-limb passes through the vertical so that it becomes overturned. *Monoclines* are another type of inclined fold, formed by a zone of steeply-dipping strata, flanked on either side by flat-lying beds. *Recumbent folds* are simply folds with axial planes dipping at less than 10°, close to the horizontal (see **155**). A cleavage is quite commonly present parallel to the axial planes of folds formed as a result of the same deformation, as shown in **153**.

Direction of Closure *Antiforms* and *synforms* can be distinguished from one another according to whether a particular fold closes upwards or downwards within its axial plane. This distinction cannot be applied to the two extremes of recumbent or vertical folds, since these folds can only close sideways within their axial planes. They are therefore known as *neutral folds*.

Tightness and Angularity The tightness of a fold is given by the angle between the fold-limbs, as measured in a direction of its axial plane. This angle gradually decreases from 180° as the fold becomes tighter, eventually becoming less than 10° in *isoclinal folds*. Another aspect of fold-style is known as *angularity*. A distinction can be made between *angular folds* with sharply defined fold-hinges, and all other folds, which generally have much more rounded hinge-zones.

Figure 153 *Upright folds with their axial planes lying close to the vertical. These folds have fold-limbs which dip in opposite directions to one another at much the same angle. There is also a crude cleavage (see* **192***) parallel to the axial planes of these folds. Alanäs, Västerbotten, Sweden.*

Figure 154 *Inclined folds affecting a banded series of quartzite and sandy dolomite, with their axial planes dipping at a moderate angle towards the right. The steeply-dipping fold-limbs are slightly overturned, so that these folds occur in the form of overfolds. Clonmass Point, County Donegal, Ireland.*

Figure 155 *Recumbent fold with a horizontal axial-plane, separating two isoclinal fold-limbs from one another. Unlike upright or inclined folds, which usually occur in the form of antiforms or synforms, recumbent folds are neutral or sidewards-closing folds. In fact, the present example closes towards the right. Svalvejukke, Västerbotten, Sweden.*

Nature of Parallel Folding

The folding of sedimentary rocks under non-metamorphic conditions commonly produces what are known as *parallel folds*. It is a characteristic feature of such folds, as shown in **156**, that the folded layers undergo little or no apparent change in thickness, as they are traced around the fold-hinges. It can then be assumed that the folded layers retain their original thickness, normal to the bedding. This implies that the original length of the folded layers may be found simply by measuring their present length around each fold-hinge in its turn. Such folds should not be termed *concentric folds*, unless the folded surfaces can be shown to occur in the form of circular arcs, with a common centre or curvature.

The geometry of parallel folds requires that they should change in shape as they are traced up and down their axial planes. In particular, the antiforms become broader and less well-defined as they are traced upwards, while they become tighter and more compressed in the opposite direction, eventually forming simple cusps in the folded surface, as shown by **157**. Synforms are exactly the opposite. This change in shape of the folded layers along the axial plane of a fold, sometimes leading to its complete disappearance, is characteristic of *disharmonic folding*.

Although there is always a disharmonic element to parallel folding, this does not mean that the deformation simply dies away beyond the limits of the folded layers. Instead, the shortening which accompanies the development of parallel folds must be expressed by deformation of some sort, affecting the surrounding rocks. This can occur in a number of different ways, depending on circumstances.

The parallel folding of highly competent layers may be accommodated by the plastic flow of a very much less competent substratum, formed for example by salt deposits. Even if such differences in competency are not so extreme, the folding of more competent layers may be accompanied by deformation, leading to the development of a slaty cleavage in the less competent layers, under appropriate conditions of temperature and confining pressure. The deformation often extends into the more competent layers as well, so modifying the nature of the folding.

Apart from such cases, the shortening undergone in response to parallel folding may be accommodated by décollement on an underlying thrust-plane. **158** shows a typical example. The surface of décollement occurs immediately below the folded horizon, where it rests on the underlying rocks, which are not affected by the folding. Although developed on a much larger scale, such surfaces of décollement often separate the folded rocks of a sedimentary cover from an underlying basement, which is likewise not affected by the movements.

Figure **156** *Parallel folds showing how the folded layers retain much the same thickness, as measured normal to the bedding, as they are traced around the fold-hinges. This style of folding may be compared with that shown by similar folds (see 176). Björkvattnet, Västerbotten, Sweden. (Field of view c 40cm)*

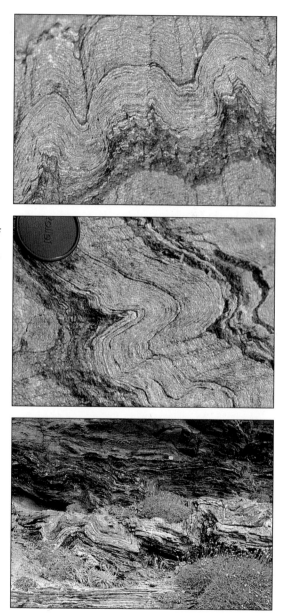

Figure **157** *Parallel folds showing the cuspate nature of the inner arcs formed in a layer that has been affected by such folding, provided that it is sufficiently intense for this feature to develop, while the outer arcs have a typically lobate outline. Björkvattnet, Västerbotten, Sweden.*

Figure **158** *Décollement affecting finely laminated rocks lying on top of a thrust-plane, which allows angular folds to form in these rocks in marked contrast to the underlying rocks, which are not folded in any way. These box-like structures are typical examples of disharmonic folds. Trebarwith Strand (SX 051867), Cornwall, England.*

Mechanisms of Parallel Folding

The parallel folding of sedimentary rocks appears to result from two distinct mechanisms, known respectively as flexural-slip folding and neutral-surface folding. Folding a telephone directory illustrates the nature of *flexural-slip folding*. The pages can be made to slip over one another to form the fold-limbs, while they are not affected by any shearing movements along the intervening fold-hinge. Likewise, flexural-slip folding occurs in the field wherever the sedimentary layers slip over one another along the bedding-planes. These movements are concentrated on the fold-limbs, dying away towards the line of the fold-hinge. However, the pages of the telephone directory must also bend in order to accommodate the folding. This occurs in the case of sedimentary layers, which have a certain thickness, as a response to *neutral-surface folding*. The bending undergone by each layer then results in extension around the outer arc of the fold-hinge, accompanied by compression along its inner arc. The *neutral surface* separates these two areas from one another. These movements die out away from the fold-hinge, so that they do not affect the fold-limbs.

Various structures can be developed as a response to flexural-slip folding.

Bedding-planes may show fault-plane striations, lying roughly at right angles to the fold-hinges, formed by flexural-slip movements on the fold-limbs. These movements result in the overlying beds being displaced up-dip, if the fold-limbs are not overturned, but down-dip wherever they are upside down. **159** shows the effect of such movements on a series of quartz veins, cutting across the bedding at a high angle.

160 shows how feather fractures may occur as another response to flexural-slip folding. The quartz vein dipping moderately to the right lies along the bedding, which can only just be recognized within the rock. A fringe of feather fractures is present along both sides of this vein, each filled with vein quartz. The principles already outlined (see **139**) indicate that the overlying rocks have moved up-dip, relative to the underlying beds. This means that the rocks are not overturned, dipping towards a synform on the right, away from an antiform farther to the left.

161 shows how *en échelon* tension gashes may also be associated with flexural-slip folds. Although forming a shear-zone, rather than a discrete surface of movement, this lies parallel to the bedding, dipping at a moderately steep angle to the right.

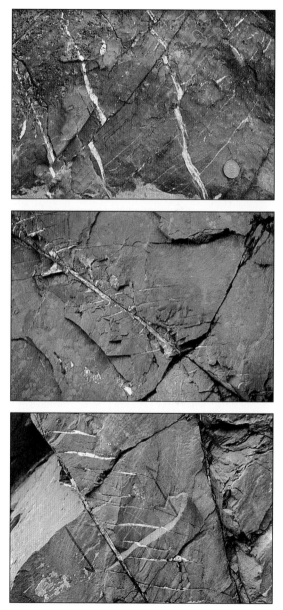

Figure 159 *Quartz veins lying at a high angle to the bedding, which have been displaced up-dip by several bedding-plane thrusts, dipping towards the left. It may be inferred that a synform lies to the left, with its complementary antiform lying somewhere to the right of this exposure. West Angle Bay (SM 853033), Dyfed, Wales.*

Figure 160 *Feather fractures forming a fringe along a bedding-plane thrust, which is itself marked by the quartz vein, dipping at a moderate angle to the right. These fractures, now filled with vein quartz, were generated by the overlying beds moving up-dip to the left, as a result of flexural-slip folding. Tebay (NY 607010), Cumbria, England.*

Figure 161 *En échelon tension gashes forming a shear-zone which is itself parallel to the bedding. The arrows drawn on the rock show how the overlying rocks have moved up the dip of the bedding, so that an antiform would be encountered towards the left, flanked by a synform to the right of this exposure. Tebay (NY 607010), Cumbria, England.*

Extensional as well as compressional features occur as a consequence of neutral-surface folding. As shown in **162**, a fringe of mineral veins may be found to occupy extension joints, lying at right angles to the bedding around the outer arc of a folded layer. There is also a suggestion that shear fractures are also present in this instance as conjugate sets, lying at a high angle to the bedding. If any displacements occur on such fractures, they would result in the formation of normal faults, which are commonly found on a large scale at least, affecting the crests of anticlinal folds. *Keystone grabens* are formed in such circumstances.

By way of contrast, compressional structures affect the cores of neutral-surface folds around their inner arcs. Often, any thin layers lying within the core of a neutral-surface fold are crumpled and otherwise folded on a small scale, as shown in **163**. This micro-folding can be seen to die away on the fold-limbs, away from the fold-hinge.

Minor thrusts may also affect the rocks lying in the core of a neutral-surface fold, cutting across the bedding at a low angle. Such thrusts often become folded at a later stage of the movements, or they may be found to break through the folded layers, so forming what has already been described as a break thrust (see **144**). Other compressional effects within the inner arc of a neutral-surface fold may lead to the local development of a slaty cleavage in suitable lithologies, while stylolites can also form in limestones, lying at a high angle to the bedding.

Fishtail and Snakehead Folds

164 shows a structure known as a *fishtail fold*. The uppermost beds are clearly arched over a bulge in the underlying rocks, even though the base of this limestone horizon is evidently not affected by these compressional movements. The individual beds of limestone penetrate one another along a whole series of minor thrusts, cutting across the bedding in a zig-zag fashion. Such interdigitation produces a structure resembling a fishtail in its form, so accounting for the name.

Snakehead folds occur in the hanging-wall rocks as a passive response to movements on major overthrusts, wherever they have a staircase geometry, as shown in Drawing 13. A hanging-wall anticline is formed in these rocks as they move up and over a ramp in the underlying thrust-plane. This produces a flat-topped structure, with a back-limb that models the form of the underlying ramp. The fore-limb, dipping in the opposite direction, represents the cut-off formed in the hanging-wall rocks where the thrust-plane cuts across the bedding to form each ramp in its turn. The resulting structure has a form much like the head of a cobra or rattle-snake about to strike, hence the name. Such hanging-wall anticlines are separated from one another by flat-bottomed synclines, reflecting the structure of the underlying thrust-plane where it follows the bedding.

Figure 162 *Extension joints, filled with vein quartz, which form a fringe around the outer arc of a folded layer. Lying at right angles to the bedding, these veins are formed as a result of neutral-surface folding, which affected the layer itself. Broad Haven (SM 860142), Dyfed, Wales. (Field of view c 3m)*

Figure 163 *Neutral-surface fold affecting a finely laminated siltstone, showing how the folded layer presents a smooth curve around its outer arc, while the very thin layers forming its fold-core are folded on a much smaller scale, as they are traced around its inner arc. Porthleven (SW 635248), Cornwall, England. (Field of view c 15cm)*

Figure 164 *Fishtail fold in bedded limestones, dipping overall at a moderate angle towards the right. A series of minor thrusts has allowed the individual beds of limestone to penetrate one another along the bedding, so forming a bulge in the overlying rocks. Scremerston (NU 025494), Northumberland, England.*

Chevron-folds and Kink-bands

The mechanisms involved in parallel folding also come into play in the case of chevron-folds and kink-bands. *Chevron-folds* have long, straight fold-limbs, separated from one another by rather angular fold-hinges. The fold-limbs need not be the same length, even though this is suggested by the name given to folds of this general type. **165** shows a series of chevron-folds with recumbent axial-planes, so that the bedding can be traced in a zig-zag fashion up the face of the cliff.

Chevron-folding usually affects well-bedded sequences of sandstone, often in the form of turbidites, where each layer has much the same thickness as all the others, separated from one another by varying amounts of shale. The relatively competent layers of sandstone then undergo neutral-surface folding in the manner of **163**, so that they each preserve what is effectively their original thickness, normal to the bedding.

However, chevron folds differ from parallel folds in that these layers maintain much the same profile at different levels within the structure as a whole. This degree of regularity can be achieved wherever the less competent rocks, separating the more competent layers from one another, are able to thicken into the fold-hinges as a result of the deformation. This effect can clearly be seen in **163**, where the dark layers are slates, formed in response to such deformation. Note how these slaty horizons are somewhat tri-angular in shape where they occupy the fold-hinges between the more competent layers of siltstone. Such a shape is dictated by the differences in curvature shown by the inner and outer arcs, formed by the boundaries of these more competent layers. If such horizons of less competent rock are lacking, cavities may open up between the more competent layers at the fold-hinges, forming *saddle-reefs* if they become mineralized.

Kink-bands are typically developed on a much smaller scale than chevron-folds, so that they can only affect finely-laminated rocks, such as slates. They usually occur in a rather sporadic fashion as discrete zones with sharply-defined boundaries, up to several centimetres apart. The slaty cleavage lying within these zones is deflected abruptly away from its position on the outside, so forming an angular fold-pair, with straight limbs and very narrow hinges. *Kink-folds* are formed wherever a whole series of kink-bands occurs in close juxtaposition with one another, as in **166**.

The geometry of kink-bands means that they can only form at an oblique angle to the external layering. It is then possible for kink-bands to occur as conjugate sets, with their axial planes lying at a high angle to one another, as shown in **167**. Such an arrangement suggests that kink-bands form parallel to planes of high shear stress in response to a direction of maximum compression, lying close to the layering itself.

Figure 165 *Chevron-folds with recumbent axial planes. Note how the bedding, which can be traced up the face of the cliff in a zig-zag fashion, forms a series of rather angular folds with relatively long fold-limbs, showing little change in profile along their axial planes. Millhook Haven (SS 186006), Cornwall, England. (Height of section c 40m)*

Figure 166 *Kink-folds affecting a slaty cleavage to form a series of very angular folds in which the fold-limbs are separated from one another by sharply-defined kink-planes. This style of folding typically occurs late in the structural history of low-grade metamorphic rocks such as slates. Onich (NN 015616), Inverness-shire, Scotland.*

Figure 167 *Conjugate kink-folds in a mylonitic rock, forming box-like structures with their axial planes dipping towards one another at a high angle. Eriboll (NC 406543), Sutherland, Scotland.*

Mechanics of Folding
Buckling

This occurs in our everyday experience wherever a mechanically rigid member of an engineering structure fails under lateral compression, undergoing a sudden deflection out of its own plane. For example, a wooden ruler compressed along its length will suddenly buckle into a sinusoidal half-wave, eventually breaking under further compression. We might expect any competent layer within a sedimentary sequence to buckle in a similar fashion as a response to lateral compression. However, the failure of this layer is constrained by the presence of the less competent material in which it is embedded, together with the weight of the overburden. Under these circumstances, **168** suggests that buckling affects such a competent layer so that it becomes deflected to form a *sinusoidal fold-train*, characterized by a particular wavelength.

Buckling may well be a mechanism which controls the development of typical neutral-surface and flexural-slip folds in sedimentary rocks. These folds often have a wavelength that depends on the thickness of the layers affected by the folding, as predicted in theory. However, the evidence that buckling is an important mechanism of folding is seen to its best advantage wherever competent layers are interbedded with less competent material, which has itself been deformed in a ductile manner under conditions of regional metamorphism.

Theory and scale-model experiments then indicate that there has to be a certain difference in competency between the layers for any buckling to occur. If this is lacking, the layers will only undergo *layer-parallel shortening* as a response to lateral compression. This would be accompanied by a gradual increase in the thickness of the layer affected in this way. However, if there are sufficient differences in competency between the layers, buckling will occur on a wavelength which depends on the relative thickness of the more competent layers. This factor is illustrated in **169**, where it can be seen that the thinner layers of light-coloured siltstone are affected by folding on a smaller scale, compared with the thicker layers. The only exception to this general rule occurs where the folding undergone by a relatively thin layer is constrained by the presence of two much thicker horizons on either side. This effect can be seen towards the centre of the field of view in **169**. Note how a thin layer of silty rock, sandwiched between two much thicker layers of siltstone with only a little dark slate in between, takes on the same wavelength as the folding which affects these layers of siltstone.

170 provides another example where it can be seen that the buckling of a particular layer generates folds with a wavelength that varies according to the thickness of the folded layers, as exemplified in this case by two sets of quartz veins.

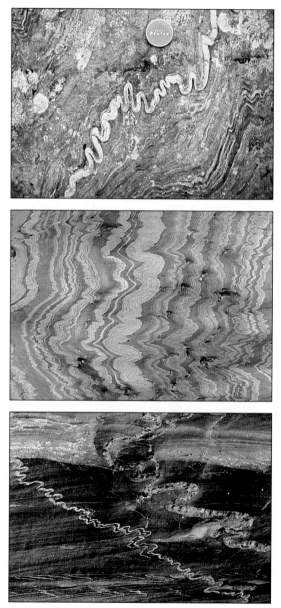

Figure 168 *Train of parallel-type folds formed by the buckling of a thin quartzite layer, surrounded by less competent rocks, which occurs as a response to lateral compression. Note how all these folds have much the same wavelength as one another. Loch Leven (169625), Inverness-shire, Scotland.*

Figure 169 *Buckle folds formed by the lateral compression of silty layers embedded in a much darker slate, showing how these folds vary in wavelength according to the thickness of the layer affected by the buckling. Porthleven (SW 635248), Cornwall, England. (Field of view c 50cm)*

Figure 170 *Folds formed by the buckling of quartz veins, differing in thickness from one another, showing how the wavelengths of these two sets of folds are directly proportional to the thickness of the quartz vein actually affected by the folding. Porthleven (SW 635248), Cornwall, England. (Field of view c 40cm)*

Buckling and Fold-style

The morphology of any folds formed as a response to buckling depends to a considerable extent on how much layer-parallel shortening has occurred, prior to the start of the folding. If there are large differences in competency between the layers, very little layer-parallel shortening will occur before the more competent layers start to buckle, probably as the result of neutral-surface folding. Flexural-slip folding may also be developed under suitable circumstances, wherever there are internal surfaces allowing flexural-slip movements to occur as a response to the buckling. This would lead initially to the development of parallel folds within the buckled layer, even though their original form might well be modified by subsequent deformation. The typical morphology shown by such folds, lacking much subsequent deformation, is illustrated in **169**.

However, many folds formed as a result of buckling do show the effects of subsequent deformation, even though there was very little layer-parallel shortening to accompany the initial stages in their development. Although clearly starting out as buckles, the folds shown in **171** have subsequently undergone a considerable degree of flattening in response to compression acting at right angles to their axial planes. This has reduced the thickness of the fold-limbs, which have become attenuated as a result, while the intervening fold-hinges have undergone a certain amount of thickening.

172 shows the typical morphology of folds formed as a result of buckling, where a certain amount of layer-parallel shortening has accompanied the initial stages in their formation. Note how the folded layers thicken into the fold-hinges, giving rise to a fold-style which departs to a considerable extent from that shown by parallel folds. This style of folding tends to occur wherever the various layers in the rock do not show very marked differences in competency between one another. Layer-parallel shortening is then favoured at the expense of buckling during the initial stages of the folding, which occurs as a somewhat belated response to the effects of lateral compression on the more competent layers. This allows the fold-hinges to thicken up preferentially as the folds gradually become tighter during the course of the deformation.

Once the folds become established in this way, layer-parallel shortening gives way to flattening, acting at right angles to the axial planes of the folds. One effect of this flattening would be to cause a thinning of the fold-limbs, once the folding had reached a certain degree of tightness, thereby reversing the layer-parallel shortening which occurred on the fold-limbs during the initial stages in the deformation. **173** shows a typical example of the fold-style which is produced as a result of this flattening. The varying degree of attenuation which affects the fold-limbs should be noted.

Figure 171 *Flattened folds formed initially in response to the buckling of more competent layers in the rock, which have then been affected by flattening, acting as a result of lateral compression at right angles to their axial planes. Eriboll (NC 414544), Sutherland, Scotland.*

Figure 172 *Flattened folds formed where layer-parallel shortening has occurred at an early stage in the buckling process, so that the surfaces defining the form of the folded layer consist of a series of cuspate fold-cores, separated from one another by more lobate fold-hinges. Rhoscolyn (SH 264749), Anglesey, Wales. (Field of view c 3m)*

Figure 173 *Flattened folds showing how the fold-limbs can be affected by differing amounts of attenuation if the deformation becomes sufficiently intense. The style of such folds provides little evidence that they were formed initially as the result of buckling, although this was probably the case. Eriboll (NC 406543), Sutherland, Scotland.*

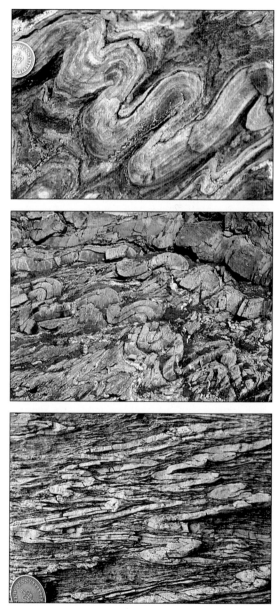

Nature of Similar Folds

Similar folds present a problem which has not yet been explained very satisfactorily by structural geologists. They are defined as any fold in which the folded surfaces have exactly the same shape as one another. This means that each layer affected by the folding, lying as it does between two such surfaces, must maintain exactly the same thickness throughout the fold, as measured in a direction parallel to the axial plane. Each layer must therefore thicken into the fold-hinges, while it shows a corresponding degree of attentuation on the fold-limbs. Such a fold preserves exactly the same profile along its axial plane, throughout all the surfaces affected by the folding.

Many folds traditionally identified as similar in style do not show the ideal geometry of perfectly similar folds, as just defined. **174** shows a typical example. Although the folding preserves much the same profile throughout the exposure, this is achieved by the competent layers buckling in a parallel fashion, while the less competent material thickens into the fold-hinges in order to accommodate the folding undergone by these layers. This association of two distinct fold-styles, restricted to different layers in the rock, which combine with one another to form similar-type folds, is characteristic of what are known as *composite similar folds*.

The similar nature of the folding then arises from the fact that the competent layers do not respond to buckling in such a way that they are entirely independent of one another. Instead, as shown in **175**, the whole assemblage of competent layers behaves as what is termed a *multilayer*. Note how the folding undergone by one layer extends along the axial planes of the folds which are formed as a result to affect the next layer, and so on, *ad infinitum*. Often, the folds formed as a result of such interference between the more competent layers in a rock have relatively straight fold-limbs, separated from one another by rather narrow hinge-zones, as for example in chevron folds (see **163** and **165**).

Any component of overall flattening, acting at right angles to the axial planes of such folds, will cause changes in their fold-style as the deformation proceeds. This tends to produce folds in the more competent layers which gradually approach the geometry of perfectly similar folds, provided that the differences in competency between the various layers are not too large (see **173**).

It is therefore most likely that the formation of perfectly similar folds, if it can be assumed that such folds actually exist, must be favoured by the virtual lack of any differences in competency between the layers affected by the folding. For example, **176** shows a metamorphic limestone in which the layering appears to reflect little else than differences in colour between the layers. The folding shows a very close approximation to the geometry of perfectly similar folds, at least in some layers.

Figure **174** *Composite similar folds in which the buckling of the more competent layers of light-coloured siltstone to form parallel-type folds is accompanied by the less competent horizons of dark slate thickening into the fold-hinges to produce folds that are overall similar in style. Porthleven (SW 635248), Cornwall, England. (Field of view c 1.2m)*

Figure **175** *Composite similar folds resulting from the folding of a whole series of quartzite layers in the form of a multilayer in which adjacent layers of more competent rock evidently do not buckle independently of one another. Loch Leven (NN 169625), Inverness-shire, Scotland.*

Figure **176** *Similar-type folds affecting a metamorphic limestone in which there appears to be very little difference in competency between the layers. The folded surfaces separating these layers from one another do not appear to differ much in shape from one another, so that the folding is close to similar in style. Knochnacur (L 932527), Connemara, Ireland.*

Disharmonic Flow-folding

It is a characteristic feature of all the mechanisms of folding so far considered that the folds formed as a result show a certain degree of regularity. This is partly a consequence of the importance which is attached to the buckling of relatively competent layers in the rock, since this tends to produce folds of a particular wavelength, assuming a constant difference in competency between these layers in the intervening material. This tendency is reinforced by the layered nature of sedimentary rocks, leading to a considerable degree of uniformity as far as their original structure is concerned. This is subsequently reflected in the nature of the folds that are produced as a response to the buckling. Furthermore, if a component of overall flattening also starts to take effect, this often acts in rather a uniform manner, so further modifying the geometry of the evolving folds in a fairly regular manner as well.

However, this regularity tends to break down in the rocks of high-grade metamorphic terrains, under conditions of high temperature at the very least, where the deformation takes on a markedly ductile character, sometimes described as plastic flow. The more competent layers may then still behave in a relatively coherent manner, particularly if they are not too far apart to buckle altogether in the form of a multi-layered sequence. However,

where these layers are isolated from one another by considerable thickness of less competent rock, the folding often takes on a disharmonic character from the very start of buckling, while the folds formed as a result tend to evolve in an erratic fashion as the deformation proceeds. This is usually expressed by the fold-limbs becoming attenuated in a very uneven fashion, eventually leading to the complete disruption of the folded layer. The end-result is a series of detached and often disorientated fold-closures, separated from one another by the plastic flow of the less competent rock.

177 shows what can be taken as the initial stages in this process, which is often marked by shearing that affects only one set of fold-limbs, causing them to become attenuated as a result. **178** shows a more advanced stage, where extreme attenuation has affected the fold-limb joining the antiform on the left to its complementary synform, lying just to its right. It is now only represented by a very thin septum of light-coloured rock, connecting together the two fold-hinges. **179** illustrates a typical example of what can best be termed *disharmonic flow-folds*, formed in response to the plastic flow of thin quartzitic layers, embedded in a much more massive mica-schist. Note how extreme thinning has affected the quartzitic layers, leaving only the fold-closures as remnants.

Figure **177** *Folds with one set of fold-limbs which has been much reduced in thickness as a result of differential movements affecting the rock-mass as a whole. Trearddur Bay (SH 251791), Anglesey, Wales. (Field of view c 1m)*

Figure **178** *Fold-pair with a highly attenuated common limb, which can be traced as a very thin septum of light-coloured rock, joining together the two fold-hinges. Note that the antiformal fold on the left has a strongly curved hinge-line, plunging towards the observer. Achmelvich Bay (NC 053260), Sutherland, Scotland. (Field of view c 2m)*

Figure **179** *Disharmonic flow-folds affecting thin quartzitic layers in a dark-coloured schist, showing how extreme attenuation of the fold-limbs has resulted in a series of isolated fold-cores, entirely surrounded by schist. Glen Orchy (NN 243321), Argyll, Scotland. (Field of view c 1.8m)*

177

Boudins and Boudinage

It is the extension undergone by a competent layer, embedded in less competent material, that leads to the formation of boudins (French: *boudin*, a sausage). The process known as *boudinage* may therefore be considered as the exact opposite to folding, which is mostly a response of layered rocks to lateral compression, acting parallel to the layering. Boudinage occurs wherever a competent layer breaks up under extension acting along its length, so forming a series of segments, which then move apart from one another to form gaps in the layer, often filled with mineral matter. **180** shows the typical appearance of boudins in cross-section, affecting a layer towards the centre of the photograph.

Where it is seen in three dimensions, a layer affected by boudinage consists of a series of elongate and rather flattened cylinders, all lying side by side so that they are parallel to one another. They then have the appearance of sausages, lying side by side on a butcher's slab, as they are displayed for sale on the continent of Europe. It was for this reason that the structure was first termed boudinage. However, most boudins are seen in cross-section, where they look like a string of short sausages, as sold in Britain and North America, all linked together end to end.

Boudins vary considerably in appearance, depending on differences in competency between the layers, the amount of deformation that has taken place, and the relative ductility of the surrounding rocks, among other factors. What can be taken as the initial stages in boudinage under relatively brittle conditions is shown in **181**. Note how the sandstone bed has undergone a certain amount of thinning to form a slight neck between its two segments, which have separated slightly to form the extension veins, now filled with fibrous quartz. Where the deformation takes place under more ductile conditions the boudins often display a cross-section, shaped somewhat like a barrel (see **180**), but usually more elongate.

182 shows a typical example of the neck which is formed between two such boudins, which have only separated from one another by a relatively short distance. Note how the layering affected by the boudinage converges on the neck separating the two boudins from one another, reflecting what must be an increasing amount of thinning away from their central parts. This feature may give a mistaken impression that a fold-closure is present at the end of each boudin, particularly if they have undergone a greater degree of separation than shown in the present example.

Figure 180 *Boudinage affecting a single layer of quartzite as a result of its extension in a direction parallel to itself. The boudins break up this layer into a series of discrete segments, each showing a barrel-shaped cross-section, separated from one another by a narrow neck of rock. Kilnaughton Bay (NR 348456), Isle of Islay, Scotland. (Field of view c 3m)*

Figure 181 *Incipient boudinage affecting a sandstone bed which has undergone brittle fracture at right angles to the bedding to form two segments, separated from one another by a slight neck, now occupied by quartz veins. Widemouth Bay (SS 197036), Cornwall, England.*

Figure 182 *Boudinage showing how a neck is formed as a result of ductile deformation affecting a zone of lateral extension, lying between two boudins. Hawks Ness (HU 459485), Shetland, Scotland.*

Scar-folds

The gaps formed between the boudins may well be the sites of mineral deposition (see **181**), or pegmatitic segregation. However, the less competent material surrounding a boudinaged layer may itself flow into these spaces as they form. As shown in **183**, this results in the formation of what is known as a *scar-fold*. The rock covered in lichen is a massive grit, which has fractured to form a series of boudins, separated from one another by much vein quartz. The mica-schist lying underneath this massive grit has squeezed into the space which opened up between these boudins as they moved apart. The hinge of such a scar-fold is parallel to the length of the boudins, lying at right angles to the direction of extension within the boudinaged layer itself.

Rotational Boudins

Most boudins appear to form simply by the extension of a relatively more competent layer, acting parallel to itself. However, boudinage is sometimes accompanied by rotational effects, so that the boudins now lie askew, at an angle to the overall attitude of the layering, as shown in **184**. It can be seen that these boudins were not formed by rupture at right angles to the boudinaged layer. Instead, they are rhomboid in shape, formed by shearing movements along fracture-planes which cut across the boudinaged layer at an oblique angle, so that they have the *appearance* of shear-fractures at the very least. It is most likely that the formation of such boudins is favoured wherever the layering has been affected by rotation away from its original position, while undergoing a certain amount of extension at the same time.

Pinch-and-swell

Boudinage does not always occur as a response to extension undergone by layered rocks, parallel to the layering. If there are only slight differences in competency between the layers, it is quite common to find that the slightly more competent layers do not separate completely, at least under relatively ductile conditions. Instead, these layers tend to show a structure known as *pinch-and-swell*, marked by the "necking-down" of the layers at more or less regular intervals along their length. A typical example is shown in **185**, where the light-coloured layers are affected in this way. It should be appreciated that the thicker parts of these layers have not undergone any actual thickening, since they have just undergone a lesser degree of thinning in comparison with the thinner portions.

Figure **183** *Scar-fold forming a deep re-entrant of mica-schist which has penetrated upwards into the space opened along a fracture in the overlying bed of massive grit as a result of its boudinage. This mica-schist moulds itself to the walls of this fracture, taking on a folded appearance as a result. Pittulie (NJ 961678), Buchan, Scotland.*

Figure **184** *Rotational boudins formed in a thin layer of quartzite as a result of its fracture along what appear to be shear-planes, lying at an oblique angle to the layering itself. Such structures may well be formed wherever the layering lies at a slightly oblique angle to the direction of extension. Clonmass Point, County Donegal, Ireland.*

Figure **185** *Pinch-and-swell affecting the light-coloured layers of more acid rock in a highly deformed gneiss in such a way that these layers apparently thicken and thin along their lengths in a fairly regular manner, forming a structure rather like a string of beads. Skjolden, Jotunhein, Norway.*

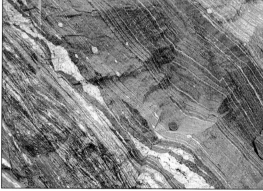

Ductile Shear-zones

As the name suggests, *ductile shear-zones* are the product of shearing movements, acting under relatively ductile conditions, which are restricted to a particular zone within a rock-body. These movements cause the rocks lying on either side of the shear-zone to slide past one another in opposite directions. They differ from fault-zones in that there is no physical break in continuity across the shear-zone, so that any faulting is entirely incidental, if present at all. The displacement across such a shear-zone is then simply a result of the ductile deformation which affects the rocks lying within its boundaries.

The nature of the deformation affecting a ductile shear-zone can be seen in **186**. This shows a quartzite in which organic burrows form a series of pipe-like structures (see also **66**). These pipes were originally at right angles to the bedding, which now dips at a low angle to the right. They are seen to preserve much the same attitude where they are present in the rather massive beds of quartzite, which are separated by a thin horizon of much less pure quartzite, seen crossing the field of view in the centre of the photograph. The clearly-defined pipes lying within this horizon have evidently been deformed by shearing movements acting parallel to the bedding in such a way that they now lie at only a moderate angle to this direction. It may then be inferred that the overlying rocks have undergone a displacement to the right, relative to the underlying rocks, which have moved in the opposite direction, as a result of the deformation that affects the intervening shear-zone. Note that the rocks lying on either side of this shear-zone are not completely lacking in deformation, since they carry pipes which are not quite perpendicular to the bedding.

Often, it is the bedding which is affected by the formation of ductile shear-zones. **187** shows a typical example, formed on a relatively small scale in well-bedded rocks. Although it is folded slightly, bedding dips steeply towards the left, while a crenulation cleavage (see **204**) is also present, dipping much less steeply in the opposite direction, parallel to the axial planes of the folds. The ductile shear-zone has itself the same orientation as this cleavage. It forms a narrow zone of highly attenuated rocks, not more than a few centimetres in thickness, which crosses the field of view from top left to bottom right. **188** provides a closer view.

The displacement occurring across this shear-zone can be determined from its effect on the bedding. If a particular bed can be traced upwards across the shear-zone from its foot-wall, it is found to reappear farther up towards the left in the hanging-wall. The hanging-wall has therefore been displaced up-dip, relative to the foot-wall. Such a displacement is typical of reverse dip-slip shear-zones, using the terminology normally applied to faults. If this displacement had occurred in the opposite direction, a normal dip-slip shear-zone would be present. The terminology can be extended to vertical shear-zones with strike-slip movements, allowing left-lateral and right-lateral displacements to be distinguished.

Figure **186** *Deformed pipes, originally forming a series of vertical burrows at right angles to the bedding, now showing the effects of shearing movements parallel to the bedding, which itself dips at a low angle towards the right. Beinn Heilam (NC 466619), Sutherland, Scotland.*

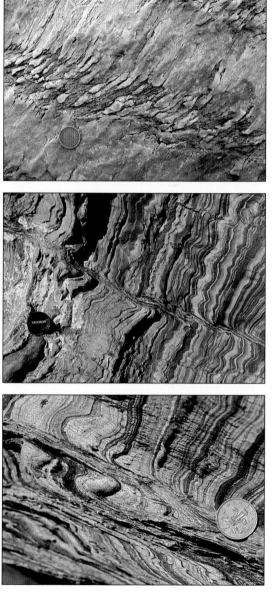

Figure **187** *Ductile shear-zone, only a few centimetres in thickness, formed by the displacement of the overlying rocks up-dip towards the left, as shown by tracing a particular bed across the shear-zone iteslf. Kintra (NR 318483), Isle of Islay, Scotland.*

Figure **188** *Closer view of the same ductile shear-zone as shown in the last photograph. Note how particular beds of light-coloured siltstone become highly attenuated as a result of the shearing movements as they are traced across the shear-zone, eventually emerging some distance away on its other side. Kintra (NR 318483), Isle of Islay, Scotland.*

The previous example showed that it was the bedding that became attenuated as a result of the displacements taking place across the shear-zone. These movements cause the bedding to rotate away from its original position outside the shear-zone, eventually becoming more or less parallel to its boundaries within the shear-zone, if these displacements become sufficiently large. Likewise, how attenuated the bedding becomes depends on the degree of displacement shown by the shear-zone, relative to its width.

Deformation fabrics may also be produced within ductile shear-zones, even though there is no earlier fabric to be affected in this way. **189** shows a ductile shear-zone cutting a coarse-grained igneous rock, lacking any internal fabric. The light-coloured phenocrysts of feldspar in this rock are evidently drawn-out by the deformation affecting the shear-zone to form a foliation, lying at an angle to its margins. Theoretically, this fabric should form at 45° to the boundaries of the shear-zone. Normally, however, the deformation gradually increases in intensity towards its centre. This is reflected by the foliation adopting a sigmoidal pattern, curving away from its 45° position at the boundaries of the shear-zone as it is traced towards the centre, where it lies at a much lower angle to this direction. The foliation would also become more intensely developed as it is traced towards the centre of the shear-zone, away from its bound-aries. Such patterns are shown by the shear-zone illustrated in **190**.

Tectonic Slides

This term was introduced into the literature to describe a fault that has a very close connection with folding under conditions of regional metamorphism. Slides typically occur on the limbs of major folds, lying at a low angle, if not parallel, to the bedding. They can be identified as *thrusts* if they are found on the overturned limbs of recumbent folds, so that they place older rocks on top of younger formations. Alternatively, they would be recognized as *lags* if they occur in exactly opposite a structural position, placing younger formations on top of older rocks as a result.

Usually, it is this evidence that certain formations are missing from the stratigraphy which allows slides to be identified in the field. However, the adjacent rocks are often more highly deformed than elsewhere, so that they have the character of ductile shear-zones along which structural discontinuities have developed as a result of the movements. **191** shows an excellent example of a tectonic slide from Horn Head in the Republic of Ireland. The flaggy nature of the rocks attests to the considerable degree of deformation that has occurred adjacent to the contact between the underlying quartzites and the dark schists, which are separated from one another by the slide itself.

Figure 189 *Ductile shear-zone affecting a coarse-grained granitic rock. Note how an oblique foliation has been produced within this shear-zone as a result of the deformation which has affected the light-coloured phenocrysts of feldspar in this granitic rock, so allowing the displacements across this shear-zone to be determined. Solvorn, Jotunheim, Norway.*

Figure 190 *Ductile shear-zone cutting across a basic igneous rock, showing how the original fabric of this rock becomes modified where it is affected by this shear-zone to form a foliation which crosses the shear-zone in a sigmoidal fashion. Tarbet (NC 160497), Sutherland, Scotland.*

Figure 191 *Tectonic slide near Horn Head in the Republic of Ireland, showing an abrupt contact between right-way-up quartzites and the overlying dark schists, which are not only inverted but older than the underlying quartzites. Mickie's Hole (B 988400), County Donegal, Ireland.*

Metamorphic Fabrics

Slaty Cleavage

This metamorphic fabric is found in any rock that has a marked tendency to split into virtually any number of very thin sheets, along planes independent of the bedding. Since these sheets are commonly used as roofing slates, any rock possessing such a fissility is known as a slate. It is typically produced under conditions of very low-grade metamorphism by the deformation of a fine-grained rock, which was originally a shale or a volcanic ash.

192 shows the typical appearance of slaty cleavage in the field, affecting bedded rocks which were originally deposited as very fine-grained volcanic ashes. The rock clearly has a tendency to split into thin slabs, parallel to the slaty cleavage. Note how weathering has accentuated the slaty cleavage, tending to open up fractures in the rock, parallel to this direction. The rocks shown in **193** are purple slates, deposited originally as shales. Bedding dips at a low angle to the right, while the slaty cleavage is again close to the vertical.

Slates are quite often cut by joints which have a curiously fluted appearance, as shown in **194**. The well-defined bedding dips moderately to the left, while the slaty cleavage dips much more steeply, so that it is more or less parallel to the direction defined by the fluting.

The fissility shown by a slate depends on the presence of an internal fabric, which examination under the polarizing microscope shows to be composed of three different elements. Firstly, the micaceous minerals in a slate always show a *preferred orientation*, so they tend to occur as parallel-sided flakes, all lying parallel to the slaty cleavage. This may arise as the result of their rotation from a more randomly orientated distribution during the course of the deformation, or it may be a consequence of recrystallization, affecting the micaceous minerals in the original rock under the influence of the stresses which produced the slaty cleavage in the first place. Secondly, quartz and any other detrital grains in a slate commonly show a *dimensional orientation*, with their long axes lying parallel to the plane of the slaty cleavage. These grains may well be deformed internally, so that they gradually change in shape as the deformation proceeds. However, it is also possible that they have been affected by *differential pressure-solution*, marked by the removal of material along the sides of each grain. This material may then be deposited in the form of *pressure-shadows*, occurring as "beards" in the plane of the slaty cleavage, at the ends of these grains. Thirdly, there is a tendency for different *domains* to form on a microscopic scale in slates, consisting of quartz-rich *microlithons*, separated from one another by micaceous seams. These compositional differences arise from the effects of pressure-solution along the micaceous seams in the rock.

Figure **192** *Slaty cleavage lying at a high angle to the bedding of very fine-grained slaty rocks, which were originally deposited as volcanic ashes. Bedding dips at a low angle towards the left, while the slaty cleavage is very close to the vertical. Beacon Hill (SK 510148), Charnwood Forest, England.*

Figure **193** *Slaty cleavage cutting across the bedding of purple slates at a high angle. Such slates can be split into very thin slabs along surfaces that follow the slaty cleavage in the rock, so allowing these slabs to be used as a roofing material. Bwlch-y-Groes (SH 563601), Gwynedd, Wales. (Height of section c 5m)*

Figure **194** *Fluted joint-surface lying at a high angle to both the bedding of a slate, and its slaty cleavage. The fine ribbing on this joint-surface is nearly if not quite parallel to the dip of the slaty cleavage, as seen within the joint-surface itself. Horton-in-Ribblesdale (SO 800702), Yorkshire, England.*

Strain-markers and the Stretching-direction

The significance of slaty cleavage can best be seen if *strain-markers* are present as objects of any known shape, subsequently affected by deformation in much the same way as the rest of the rock. **195** shows a slaty rock formed from a volcanic ash containing *accretionary lapilli*. Although originally circular in cross-section, they are now seen to be elliptical objects, flattened in the plane of the slaty cleavage. **196** shows a surface formed by splitting the same rock along the slaty cleavage. Bedding is inclined slightly towards the left. Note in comparison with **195** that the lapilli show very much less deformation in the plane of the slaty cleavage, than at right angles to this direction. However, they are all slightly elongated parallel to one another at a high angle to the bedding.

Similar relationships are found whatever type of strain-marker is present in a slaty rock. Slaty cleavage therefore appears to form by *flattening* at right angles to a direction of maximum compression in the rock, accompanied by a certain degree of *stretching* within the plane of the slaty cleavage, at right angles to this direction. The stretching occurs parallel to a direction of maximum extension in the rock.

It is often very difficult to assess what changes in volume have occurred during the development of a slaty cleavage. There would be a reduction in volume if the cleavage started to form before all the pore fluids were expelled from the rock during its compaction. Pressure-solution would have a similar effect, wherever the material removed in solution is completely lost to the system, without being replaced by any other material from elsewhere.

The Stretching-direction The flattening undergone by deformed objects may result in their extension, acting uniformly in all directions within the plane of a slaty cleavage. More commonly, this extension occurs preferentially in a single direction, plunging steeply down the dip of the slaty cleavage in most cases. A fibrous mineral lineation, known loosely as the *stretching-direction*, may then be developed. It is often accentuated by the presence of deformed objects, which are all elongated parallel to one another in this direction. **197** shows a series of deformed blebs of pyrite, all lying within a slaty cleavage with their long axes defining the direction of stretching within the rock.

Pressure shadows often form at the ends of particularly rigid objects, where they occur as fringes of recrystallized material, mostly quartz and chlorite, elongated parallel to the stretching-direction. Alternatively, such objects may break up into discrete fragments, so allowing fibrous minerals like quartz and chlorite to grow in the spaces between these fragments, where they again occur parallel to the stretching-direction.

Figure **195** *Accretionary lapilli in a "bird's eye tuff" forming strain-markers, strongly flattened within the slaty cleavage which is itself horizontal. Kentmere (NY 459074), Cumbria, England. (Field of view c 5cm)*

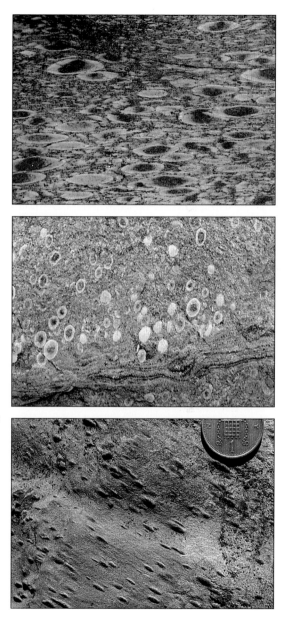

Figure **196** *Accretionary lapilli from the same locality as the previous photograph, as seen on a surface parallel to the slaty cleavage, rather than at right angles to this direction. There has been only a slight amount of stretching within the slaty cleavage, so that the deformation was mostly a response to flattening. Kentmere (NY 459074), Cumbria, England. (Field of view c 6cm)*

Figure **197** *Deformed blebs of pyrite, all lying within a slaty cleavage with their long axes parallel to one another, defining the stretching-direction within the rock. This direction is also parallel to a fibrous mineral lineation which can be seen on the surfaces formed by this slaty cleavage. Exact locality not known, Argyll, Scotland.*

189

Penetrative Deformation and Spaced Cleavages

Slaty cleavage is commonly present as a very pervasive structure, which penetrates the whole rock-mass to produce what is known as a *penetrative cleavage*. The rock can then be split along an almost infinite number of planes, all lying parallel to one another. However, this uniformity shown by a slaty cleavage is partly a matter of scale. Cleavages which appear to be penetrative if seen in hand-specimen, even using a magnifying glass, may be found to consist of quartz-rich *microlithons*, separated from one another by micaceous seams, when examined under the microscope.

This feature can hardly occur in slates which are composed almost entirely of micaceous material, without any quartz or other detrital minerals. It becomes more marked, therefore, as the original sediment becomes coarser in grain-size, while coming to affect more of the rock as the deformation increases in intensity. This often forms a new banding in the rock, marked by differences in composition, which eventually forms what is known as a *spaced cleavage*.

Slaty cleavage is often associated with what appear at first sight to be shearplanes in the rock, which may themselves be a form of spaced cleavage. **198** shows a typical example where the bedding is clearly offset along a series of closely-spaced planes. Such offsets may be a consequence of pressure-solution, if it has the effect of removing very substantial amounts of material along the trace of each plane in the rock. However, it is difficult to apply this theory to the present example, without inferring that some shearing movements have occurred along these planes, since the bedding lies at right angles to their direction. Offsets would in fact only occur as a result of pressure-solution if the bedding lay at an oblique angle to these planes of spaced cleavage. **199** shows another example, which provides rather better evidence for the role played by pressure-solution in the development of slaty cleavage.

The exact nature of a spaced cleavage varies to a considerable extent according to the lithology of the original rock. **200** shows a typical example, which might have been termed a *fracture cleavage* in the past, although this is now considered to be an obsolete term, owing to its genetic connotations. Although the microlithons lying between the planes formed by a spaced cleavage need not be affected by the deformation, they are often cut by a penetrative cleavage, lying parallel to the spaced cleavage, or sometimes at an oblique angle to this direction. Where an earlier cleavage is present, folded in such a way that it traces out a sigmoidal path as it crosses these microlithons, such a spaced cleavage passes into what is known as a *crenulation cleavage* (see **204**).

Figure **198** *Spaced cleavage in a slaty rock, forming a series of closely-spaced planes, across which the bedding has been displaced by a small amount, apparently as a result of shearing movements. Rhoscolyn (SH 264749), Anglesey, Wales.*

Figure **199** *Morphological detail shown by a spaced cleavage-plane, suggesting that pressure solution has been an important element in its formation. Rhoscolyn (SH 264749), Anglesey, Wales.*

Figure **200** *Spaced cleavage preferentially developed in finer-grained sediment lying below a bed of massive sandstone. It forms a closely-spaced parting in the rock, dipping towards the right at a much steeper angle than the bedding itself. Cwmtudu (SN 355576), Dyfed, Wales.*

Refraction and Fanning of Cleavages

The refraction of a slaty cleavage typically occurs wherever it changes systematically in attitude as it passes through a series of beds, differing from one another in lithology, as shown in **201**. The darker beds are slates, in which the slaty cleavage dips much more steeply towards the right than the bedding. However, this cleavage in entering the light-coloured beds of volcanic ash, which are coarser in grain-size and therefore more competent, comes to lie at a much higher angle to the bedding, as a result of its refraction. Note that this occurs in a rather abrupt fashion as the cleavage passes into the prominent bed of light-coloured rock from below, while it shows a more gradual change in attitude as it crosses this bed into the overlying slates. This allows the presence of graded bedding to be recognized, showing that the rocks are not inverted in the present instance. In fact, the base of this bed is coarser-grained than its top, while it rests with a sharp contact on the underlying slate.

Development of Cleavage Fans Slaty cleavage often occurs parallel to the axial planes of any folds with which it is associated. However, such a relationship only holds good along the axial plane itself, if the cleavage is affected by refraction on the fold-limbs. *Cleavage-fans* are then developed in the more competent layers, as shown in **202**. Note how the cleavage-planes change in attitude around the fold-hinge in such a manner that they converge on one another as they are traced from the outer arc of this fold, towards its core. Such a *convergent cleavage-fan* is typically shown by spaced cleavages, where they affect the more competent layers.

A less competent layer generally shows the planes of slaty cleavage as diverging from one another if they are traced in the same direction, so forming a *divergent cleavage-fan*. **203** shows a thin band of volcanic ash, which is intensely folded, while the surrounding slates are affected by a well-developed cleavage. Divergent cleavage-fans can be seen quite clearly where this cleavage isolates triangular-shaped areas of less-deformed rock, which form "strain-shadows" around the outer arcs of particular folds within this ash band.

A very unusual form of divergent cleavage-fan is found in rocks which show the presence of what is known as a *finite neutral point*. This is a point of no deformation along the axial plane of a fold, where extension around the outer arc of a neutral-surface fold gives way to layer-parallel shortening at right angles to the plane of the slaty cleavage. A cleavage is then found very locally parallel to the bedding as it is traced around the outer arc of such a fold, flanked on either side by curved cleavages to form a triangular-shaped area with curved sides around the finite neutral point itself.

Figure **201** *Refraction of a slaty cleavage in passing through a graded bed of volcanic ash, dipping towards the right. Note how this slaty cleavage shows an abrupt change in attitude as it crosses the base of this graded bed, whereas it forms a smooth curve as it passes through this bed into the overlying slates. Bradgate Park (SK 529112), Charnwood Forest, England.*

Figure **202** *Fanning of a spaced cleavage, present in a prominent series of greywacke beds, as these beds are traced around a synformal hinge. The near-vertical dip shown by these cleavage-planes on the gently-inclined limb of this fold changes into a moderate dip towards the right on its other near-vertical limb. Cwmtudu (SN 359580), Dyfed, Wales.*

Figure **203** *Divergent cleavage-fans associated with the ptygmatic folding of a thin band of volcanic ash in a slate, forming triangular areas of less-deformed rock around the outer arcs of these folds. Longcarrow Cove (SW 893768), Cornwall, England. (Photograph by C.T. Scrutton)*

Crenulation Cleavage

The characteristic features shown by a *crenulation cleavage* arise from the micro-folding of a pre-existing fabric in the rock. Although the bedding may occasionally be affected, it is much more commonly found to be a slaty cleavage, dating from an earlier phase in the deformation-history. The initial stages usually cause finely-marked crenulations to affect the slaty-cleavage, as shown in **204**, forming very small microfolds on its surface, often with rather sharp hinges.

If a crenulation cleavage affects a slaty cleavage which is exceedingly fine-grained, the microfolding then occurs on a correspondingly small scale. This produces a crenulation cleavage which often cannot be clearly distinguished from a slaty cleavage in the field, without the use of a hand-lens at the very least. **205** illustrates this particular difficulty. It shows a crenulation cleavage that is developed on a very fine scale as an *axial-planar cleavage* to folds of the bedding. Note the spaced character, as marked by the slight striping which is shown by the darker rocks, parallel to this cleavage. No fabric is visible within these stripes to show the presence of an earlier cleavage. However, by examining the bedding where it is folded around the fold-hinges, it may be possible to recognize this earlier fabric, crenulated somewhat in the manner of **204**.

The nature of a crenulation cleavage can best be observed where it affects a coarse-grained schistosity rather than a slaty cleavage, particularly where a lithological layering is also present in the rock. **206** then shows how it typically occurs as a spaced cleavage, forming discrete planes along which the rock has a tendency to split. These planes may simply be formed by the axial planes of any microfolds which affect the earlier fabric, particularly where this fabric is folded around a later fold-hinge. However, a crenulation cleavage is usually found lying at an oblique angle to the earlier fabric. The planes of crenulation cleavage then mark the attenuated limbs of these microfolds, where they have been rendered mica-rich through the selective loss of quartz, presumably as a result of pressure-solution. This can be seen clearly on the left-hand side of **206**, where the earlier fabric takes on a sigmoidal form as it crosses the micro-lithons between these cleavage-planes.

Although the layering can be traced without any break from one microlithon to the next in some places, elsewhere it is sharply offset across the intervening planes of crenulation cleavage. The latter relationship is particularly clear on the right-hand side of **206**. Although it is now considered likely to be a consequence of pressure-solution, rather than shearing movements, this gave rise to the now-obsolete name of *strain-slip cleavage*, which was previously applied to this form of spaced cleavage.

Figure 204 *Crenulations affecting a surface formed by a plane of slaty cleavage, caused by the microfolding of this surface during a later stage in the deformation history which first produced the slaty cleavage in this rock. The rather angular hinges of these crenulations define a lineation within this surface. Easdale (NM 748171), Argyll, Scotland. (Field of view c 5cm)*

Figure 205 *Crenulation cleavage in a very fine-grained slate, showing how such a cleavage may be very difficult to recognise in the field, even if a hand-lens is used. The slight striping shown by this cleavage, axial-planar to folds of the bedding, is the only clear indication of its spaced nature, Porthleven (SW 635248), Cornwall, England. (Field of view c 25cm)*

Figure 206 *Crenulation cleavage developed on a much larger scale than is usually the case, affecting the layering in a coarse-grained schist to form a series of microlithons, separated from one another by mica-rich areas. Note how the layering can be traced in a sigmoidal fashion across these microlithons. Björkvattnet, Västerbotten, Sweden. (Field of view c 45cm)*

Strain-bands

Strain-bands typically occur in well-foliated rocks, formed by the intense development of a spaced cleavage as well as a penetrative schistosity. This produces a compositional banding on a very fine scale, which is then folded between parallel sets of closely-spaced planes, as a response to a later phase of deformation. The resulting structure, as shown in **207**, has some features in common with kink-bands. However, the folding is usually rather less angular, since the layering is deflected to some extent in a sigmoidal fashion as it crosses each strain-band in its turn.

Another difference is illustrated in **207**, where it can be seen that the layering within the strain-bands lies at a higher angle to their boundaries, while the layering on the outside makes a lower angle with this direction. Note also how the layering becomes more finely-banded, and apparently less rich in quartz, as it is traced away from each strain-band into the surrounding rocks. This suggests that pressure-solution is likely to be an active process in the formation of strain-bands, affecting the layering on their long limbs.

All these features suggest that strain-bands are more closely related to the development of crenulation cleavages, rather than kink-bands. In fact, these two types of structure are often found in close association with one another, formed as a response to the same phase of deformation by slightly different rock-types. A gradual transition in structural style can often be demonstrated in the field, as strain-bands pass into what is effectively a crenulation cleavage, although present on a much larger scale, as they become more closely-spaced in response to increasing amounts of deformation. **208** shows the typical appearance of such a cleavage where it occurs within a rather complex strain-band.

209 shows two features which are commonly seen in strain-bands, particularly where they form the dominant structures in a rock as the result of fairly intense deformation. Firstly, the layering becomes highly attenuated as it crosses the boundaries of the strain-bands into the surrounding rocks, at least in comparison with its appearance elsewhere within these strain-bands. This may be the result of high shear strains, caused by shearing movements directed along the boundaries between the strain-bands, and affecting the intervening rocks. However, it is equally possible that a very considerable amount of pressure-solution has also occurred, affecting the layering only where it lies outside the strain-bands. Secondly, the layering lying within the strain-bands is quite clearly affected by buckles, which lie with their axial planes at an angle to the boundaries of the strain-bands. This can best be seen towards the centre of the field of view. Evidently, this has occurred in response to compression, acting in a direction close to the vertical in the present case.

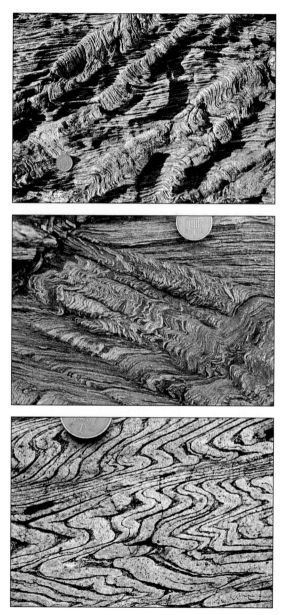

Figure **207** *Strain-bands affecting an earlier foliation which has been folded to form what appears to be parallel-sided shear-zones in the rock, across which the layering can be traced in a sigmoidal fashion. Trearddur Bay (SH 249792), Anglesey, Wales.*

Figure **208** *Closely-packed strain-bands affecting an earlier foliation which have more the appearance of a crenulation cleavage on a large scale. Note how the strain-bands themselves form microlithons of relatively quartz-rich rock, while the folded layering becomes somewhat attenuated as it crosses the intervening areas. Trearddur Bay (SH 249792), Anglesey, Wales.*

Figure **209** *Complex strain-bands showing how the layering becomes highly attenuated as it crosses from one microlithon to the next, so that the microlithons themselves form a secondary layering in the rock. Note also the presence of buckles within these microlithons with their axial planes lying at a distinct angle to their boundaries. Rotmell (NO 013473), Perthshire, Scotland.*

Part v
Structural Relationships in Folded Rocks

Strain-band showing extreme attenuation of the layering away from its boundaries. Rotmell, Scotland.

STRUCTURAL SCALE AND FOLD GEOMETRY

The structures produced by deformation in regionally metamorphosed rocks can be considered to occur on two distinct scales, if we ignore for our present purposes any structural features that can only be seen under the microscope. *Minor structures* are defined as all those features that can be observed on a small scale within the confines of a single exposure, without using a hand-lens. They typically include minor folds, cleavages and foliation, and various types of lineation. By way of contrast, *major structures* cannot be observed directly in the field since they occur on too large a scale, unless the rocks are almost completely exposed over a sufficiently wide area, as might be formed by a cliff-face or the side of a mountain. Lacking such evidence, we usually need to consider how the minor structures are related to one another at each exposure in order to determine what major structures are present within a particular area. Mapping out the major structures in this way requires us to understand in some detail the geometrical relationships which exist between minor structures of various types. This is best done by considering the geometry of what is defined strictly as cylindrical folding.

Any surface is said to be folded cylindrically if it has a form which could be generated exactly by a straight line moving parallel to itself around a curved path (see **150**). This straight line is known in very general terms as the *fold-axis*. It is parallel to any fold-hinges that are present within the folded surface. Although a surface affected by cylindrical folding obviously varies in attitude as it is traced around a particular fold-hinge, there is always a line lying within this folded surface that maintains a constant direction, parallel to the fold-axis, as shown in Drawing 15. This line therefore corresponds to the plunge of the folded surface. There are two important consequences that arise from this geometrical feature as far as cylindrical folds are concerned. Firstly, any minor folds present within a cylindrically folded surface must plunge in exactly the same direction as the major folds, as shown in Drawing 16. If this is not the case, the surface can no longer be folded in a cylindrical manner. Secondly, any axial-planar cleavage must intersect the cylindrically folded surface in a direction that corresponds exactly to the plunge of the fold-axis, as shown in Drawing 17. Similar relationships are seen in cylindroidal folds except that all these various directions can only be considered as lying approximately parallel to one another, given that cylindroidal folding is taken as only an approximation to cylindrical folding.

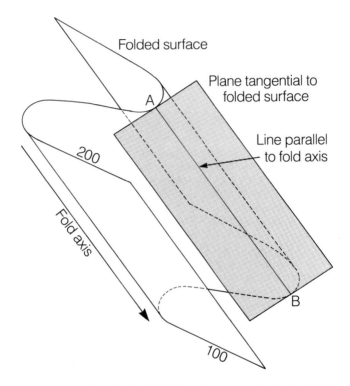

Drawing **15** *Geometry of a cylindrically folded surface, generated by the straight line AB moving parallel to itself around a curved path. Note how any planar element of such a folded surface always contains a line parallel to the plunge of the folds. (Roberts, Fig. 4.3A,* Introduction to Geological Maps and Structures, *published by permission of Pergamon Press.)*

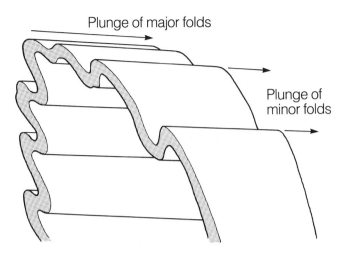

Drawing 16 *Any minor folds in a cylindrically folded surface have the same plunge as the major folds. (Roberts, Fig. 8.9,* Introduction to Geological Maps and Structures, *published by permission of Pergamon Press.)*

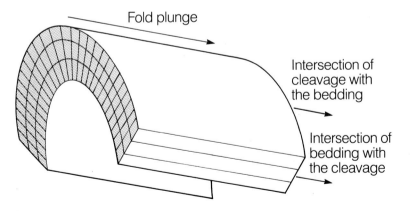

Drawing 17 *How to determine the plunge of a major fold from the direction in which bedding and an axial-planar cleavage intersect one another, assuming the folding to be cylindrical in nature. (Roberts, Fig. 8.7,* Introduction to Geological Maps and Structures, *published by permission of Pergamon Press.)*

Major and Minor Folds

Minor folds occur on a sufficiently small scale that they can be observed as structural entities within the confines of a single exposure. Typically, they have the form of *parasitic folds*, found on the fold-limbs and around the hinge-zone of any larger fold. Although these structures have been termed *drag-folds* in the past, this term implies that they formed as a result of frictional drag between more competent layers during the course of flexural-slip folding. If so, minor folds should only be found on the limbs of major folds. However, since they are frequently found around the major hinge-zones as well, it may be inferred that minor folds are formed as a consequence of buckling, which generates folds differing from one another in wavelength. **210** shows how folding on different scales can then affect a single layer, formed by a quartz vein.

Minor folds are said to be *congruent* with a set of major folds if they all share the same axial-planar cleavage. This places important constraints on the form of the minor folds, which can then be used to map out the major folds. This is particularly clear at the major fold-hinges, where the minor folds must have limbs of roughly the same length, since the layering lies overall at right angles to the axial plane of the major fold at this point. This means that the minor folds are symmetrical in profile, shaped like an M or a W, when viewed down-plunge as shown in **210**.

Traced away from a major fold-hinge, minor folds will show an increasing sense of asymmetry, with fold-limbs that differ from one another in length. Looking down-plunge, this sense of asymmetry can be designated as clockwise or anti-clockwise, according to whether the minor fold-pair, joined together by a short limb in common with one another, are shaped like a Z or an S, respectively.

For example, **211** shows a Z-shaped minor fold-pair. Since the axial planes dip moderately towards the left, any major fold is likely to have the same inclination. It may then be inferred that a major synform would be encountered at a higher structural level towards the left, whereas a complementary antiform would be present at a lower structural level in the opposite direction.

In contrast, **212** shows an S-shaped minor fold-pair. The axial planes are close to recumbent, showing a lower angle of dip than the bedding in the long limbs. This means that this minor fold-pair occurs on the overturned limb, lying between a major recumbent fold developed at a higher structural level towards the right, and another recumbent fold, which would be found at a lower structural level towards the left.

Figure 210 *Folds formed by the buckling of a quartz vein, showing how folding can occur on two distinct scales to generate a series of minor folds, which are themselves superimposed on top of another set of folds affecting the quartz vein on a larger scale, so forming a set of major folds. Loch Monar (SH 197388), Inverness-shire, Scotland.*

Figure 211 *Minor fold-pair forming Z-shaped folds with a clockwise sense of asymmetry, developed on the normal fold-limb between a major synform lying down-dip to the left, and a major antiform lying up-dip to the right. Both these major folds are likely to be inclined structures with their axial planes dipping moderately to the left. Eriboll (NC 414544), Sutherland, Scotland.*

Figure 212 *Minor fold-pair forming S-shaped folds with an anti-clockwise sense of asymmetry, lying on the overturned fold-limb between a major recumbent fold, closing up-dip to the right at a higher structural level, and another recumbent fold, closing down-dip to the left at a lower structural level. Cracklington Haven (SX 139966), Cornwall, England. (Field of view c 16m)*

Bedding–cleavage Intersections

The common occurrence of slaty and other cleavages parallel to the axial planes of any associated folds means that the relationship of bedding to the cleavage can be used to determine the nature of the folding on a larger scale. Firstly, there is obviously a direct correspondence between the dip and strike of such a cleavage and the inclination shown by the axial planes of these folds. Even if cleavage-fans are present, its attitude can be measured at the fold-hinges where it would be parallel to the axial plane of a particular fold.

Secondly, the plunge of a cylindroidal fold can be determined from the intersection of an axial-planar cleavage with the bedding, or whatever surface is affected by the folding. This arises from the fact that the axial plane is defined as a surface which separates the two fold-limbs from one another. The axial plane therefore intersects each bedding-plane in a line, coinciding with the fold-hinge which lies within that particular surface (see **150**). Such a relationship extends throughout all the rocks affected by the folding, provided that its cylindroidal character is also maintained throughout this area. This means in effect that any intersections formed by the bedding with an axial-planar cleavage will be parallel to one another, not only at the fold-hinge but also on the fold-limbs of a cylindroidal fold, whatever the scale of the folding.

213 shows the typical appearance of such an intersection of the bedding with an axial-planar cleavage. The upper surface of the exposure is formed by a flat-lying plane of slaty cleavage, while the lithological layering defines the bedding, close to the vertical. The slaty cleavage then intersects the bedding in a direction which is parallel to the yellow marker, placed on the slaty cleavage. This direction would be parallel to the plunge of any associated folds.

Another example of such a bedding–cleavage intersection is shown in **214**, where the surface facing the camera is formed by a steeply-inclined bedding-plane. A fine ribbing can be seen on this surface, plunging steeply towards the right, down the dip of this bedding-plane. This is formed by the intersection of a spaced cleavage with the bedding-plane. Similar features are commonly seen on the surface formed by a bedding-plane where it is intersected by a slaty cleavage.

Taken together, these two examples demonstrate that bedding–cleavage intersections are expressed by differences in lithology when observed on axial-planar cleavages, whereas they are simply seen as structural features where they occur on the bedding (see also Drawing 17). Similar intersections may also be produced by crenulation cleavages, as shown in **215**, where a fine crenulation can be seen on a vertical bedding-plane, plunging at a low angle towards the right. Close examination would show that it is formed by the microfolding of an earlier fabric.

Figure **213** *Lineation formed by the intersection of bedding with a slaty cleavage. Note how the bedding produces a lithological striping on the surface formed by the slaty cleavage. Bwlch-y-Groes (SH 563601), Gwynedd, Wales.*

Figure **214** *Lineation formed by the intersection of a spaced cleavage with a bedding-plane, forming the surface that dips steeply towards the camera. Note how the cleavage produces a fine ribbing on this surface, rather than a lithological striping as seen in the previous photograph. Kilmory Bay (NR 741700), Argyll, Scotland. (Field of view c 1.8m)*

Figure **215** *Lineation formed by the intersection of a crenulation cleavage with a bedding-plane, forming a vertical surface facing the camera. It forms a fine crenulation, produced by the micro-folding of an earlier fabric in the rock, such as a slaty cleavage. Kirkdale (NX 518539), Kirkcudbrightshire, Scotland. (Field of view c 1.2m)*

Bedding–cleavage Relations

The intersection of the bedding with an axial-planar cleavage gives the plunge of any folds that are related to this cleavage, assuming that the folding is cylindroidal. The relationship of the bedding to this cleavage can then be examined within the plane of the fold-profile, simply by looking down the plunge of their intersection with one another.

Where an axial-planar cleavage dips at a steeper angle than the bedding, it can be inferred that the bedding has not been structurally overturned as a result of the folding. It then lies on a fold-limb which dips towards a synform in one direction, away from an antiform in the opposite direction. It is important to realize that such a structural relationship tells us nothing about the stratigraphic order of the beds affected by the folding, contrary to the statements often made in elementary text-books. This can usually only be determined if there is independent evidence from the study of sedimentary structures as indices of stratigraphic order to show whether or not the bedding has been affected by stratigraphic inversion.

For example, **216** shows a series of upright folds where a spaced cleavage dips much more steeply than the bedding, parallel to their axial planes. Graded bedding in these rocks provides independent evidence that they are right-way-up. However, if the photograph is turned upside down, the structural relationships still remain the same, except that the antiform previously seen in the centre of the field of view now appears to be a synform, while the beds would obviously be inverted. Note, however, that the spaced cleavage still dips at a steeper angle than the bedding even under these circumstances of stratigraphic inversion.

Bedding–cleavage relationships can therefore only be used to demonstrate that structural overturning has occurred on the inverted limbs of major over-folds. **217** shows a typical exposure, where a spaced cleavage dips towards the left at a lower angle than the bedding. If this cleavage is associated with any major folds, an antiform would be encountered towards the left at a higher structural level, while its complementary synform would be found in the opposite direction at a lower structural level.

218 provides another example of the same relationship, where a slaty cleavage dips towards the left at a somewhat lower angle than the bedding. Similar conclusions can be drawn from the nature of this exposure, as made in the previous example. However, it also demonstrates that the angle made between an axial-planar cleavage and the bedding on the limbs of a major fold becomes increasingly small, as the folding becomes progressively tighter. Very careful observation is therefore needed to distinguish an axial-planar cleavage from the bedding, especially when dealing with isoclinal folds, where this technique of structural mapping is particularly useful.

Figure 216 *Spaced cleavage affecting a series of sandstone beds, separated from one another by slaty horizons. This cleavage is axial-planar to the upright folds of the bedding, despite the fact that it shows a certain amount of fanning in a rather irregular way as it is traced around the fold-hinges. South Stack (SH 205823), Anglesey, Wales. (Field of view c 15 m)*

Figure 217 *Spaced cleavage in a cleaved siltstone, dipping towards the left at a lower angle than the bedding, showing that this exposure lies on the overturned limb between two major fold-closures, consisting of an antiform to the left and a synform to the right. Kilmory Bay (NR 697748), Argyll, Scotland.*

Figure 218 *Relationship of a slaty cleavage to the bedding, showing evidence of structural inversion from the fact that the slaty cleavage dips towards the left at a slightly lower angle than the bedding, which must therefore be overturned. Cracklington Haven (SX 139968), Cornwall, England.*

Anticlines, Synclines and Structural Facing

It is the use of sedimentary structures as indices of stratigraphic younging which allows a particular fold to be recognized as an anticline or a syncline, as the case may be. For example, the recumbent fold shown in **219** is a syncline with a core of younger rocks. This can only be demonstrated in the field from the presence of graded bedding, which shows the upper fold-limb to be stratigraphically inverted. However, if graded bedding were to show that the bedding was inverted on the lower limb of this fold, not its upper limb as just stated, this fold would then be an anticline, with a core of older rocks. This demonstrates quite clearly that it is the stratigraphic order of the beds which determines whether a particular fold occurs as an anticline or a syncline, not its geometrical form.

Once this is appreciated, the concept of what is known as *structural facing* can be introduced. This is defined by the direction in which progressively younger beds are encountered along the axial plane of any fold-structure, at right angles to its hinge-line. The syncline shown in **219** can then be described as facing horizontally towards the right within its axial plane. However, if this fold were in fact an anticline, with a core of older rocks, it would then be found to face in the opposite direction within its axial plane, towards the left. The direction of facing shown by a particular fold-structure therefore depends on whether it forms an anticline or a syncline, whatever its geometrical form.

The direction of structural facing can therefore only be determined if there are sedimentary structures present to act as indices of stratigraphic younging. The necessary observations can be made at any suitable fold-closure. For example, **220** shows a recumbent fold, closing towards the right, which affects a graded bed of fine-grained greywacke and siltstone. The grading in this bed clearly indicates that the bedding is inverted on the lower limb of this recumbent fold. It therefore forms an anticlinal closure, with a core of older rocks, which faces horizontally towards the right, parallel to the axial-planar cleavage developed in the finer-grained beds.

More commonly, the direction of structural facing can be determined on any exposed surface formed by an axial-planar cleavage. As shown in **221**, the intersection of the bedding with such a cleavage-plane defines the plunge of the associated folds. It also provides a cross-section through any sedimentary structures that are present within the bedding. It can be seen in the present instance that the thicker beds of dark sandstone are graded, showing rather abrupt contacts with the underlying rocks. The fold-structures associated with this cleavage then face upwards within this cleavage-plane at right angles to the direction of its intersection with the bedding.

Figure **219** *Recumbent fold forming a synclinal closure in schistose grits. Since graded bedding can be used in the field to show that the bedding on the upper limb of this fold is inverted, its core must be formed by younger rocks, so demonstrating that this fold is a syncline rather than an anticline. Collieston (NK 060308), Buchan, Scotland. (Field of view c 10m)*

Figure **220** *Anticlinal fold-closure with a core of older rocks, as seen from the presence of graded bedding on the lower limb of this recumbent fold, which shows this fold-limb to be upside-down. This fold therefore faces to the right in the direction of its axial plane. Millook Haven (SS 186006), Cornwall, England.*

Figure **221** *Intersection of the bedding with a slaty cleavage, showing the associated fold-structures to face upwards within this axial-planar cleavage from the presence of graded beds of dark sandstone, each showing a rather abrupt contact with the underlying rocks. Bwlch-y-Groes (SH 563601), Gwynedd, Wales.*

Lineations in Folded Rocks

Minor folding is often accompanied by structural features of a linear nature, lying parallel to the direction of the local fold-hinges. Such structures come under the general heading of *lineations*, which have very diverse origins.

Some lineations are formed by the intersection of two planes in a rock, when they are known as *intersection-lineations*. The intersection of bedding with a slaty cleavage (see **213**) would constitute such a lineation, lying parallel to the hinge-lines of any associated folds. If there is a pre-existing fabric in the rock, perhaps formed as a result of compaction at right angles to the bedding, *pencil structure* may be produced if the subsequent deformation is relatively weak. The rock then breaks into pencil-like fragments, often along rather irregular planes, parallel to the local fold-hinges.

The microfolding of an earlier fabric may be associated with the intersection of two cleavages in a rock, as where a crenulation cleavage affects a slaty cleavage (see **215**). This produces what is generally known as a *crenulation-lineation*, even although it is really just an intersection-lineation in a different guise. **222** shows a surface formed by a plane of slaty cleavage, on which two very distinct lineations can be seen. The earlier lineation, plunging very steeply towards the right, marks the intersection of the bedding with this plane of the slaty cleavage. The second lineation, plunging towards the right at only a moderate angle, represents the intersection of a crenulation cleavage with the same plane.

Analogous structures are seen in metamorphic rocks of a much higher grade, where bedding and an earlier schistosity are often found to be affected by a later schistosity, axial-planar to a later set of minor folds. Their intersection often produces a coarse-grained lineation, often somewhat bladed in its character, as shown in **223**.

More penetrative deformation under conditions of increasing metamorphic grade often leads to the development of what can be termed a *grain-fabric lineation*, perhaps best seen in rocks rich in quartz and feldspar, which is defined by the shape, size and mutual arrangement of the various mineral grains in the rock. The exact nature of such a lineation, as shown in **224**, is often difficult to discern in the field, without the use of a microscope, and it is not always parallel to the local fold-hinges. A variant is produced wherever any acicular minerals in a rock all lie with their long axes parallel to one another to form what is more usually known as a *mineral lineation*. Alternatively, pebbles and other objects of a discrete nature may be found in highly deformed rocks, all lying with their long axes parallel to one another, so forming a *stretched-pebble lineation*.

Figure **222** *Intersection-lineations occurring as two sets on a surface formed by a slaty cleavage. The bedding intersects this surface to form an earlier lineation, plunging steeply towards the right, while a crenulation cleavage forms a later intersection, plunging much less steeply in the same direction. Kintra (NR 318483), Isle of Islay, Scotland.*

Figure **223** *Intersection-lineation in a coarse-grained gneiss, plunging at a low angle to the right, parallel to a set of fold-hinges in this rock. Achmelvich Bay (NC 053260), Sutherland, Scotland.*

Figure **224** *Grain-fabric lineation developed in a highly-deformed rock as a reflection of the dimensional orientation shown by the various minerals that make up this rock. This lineation lies parallel to a set of isoclinal fold-hinges. Loose block, Strathtongue (NC 618597), Sutherland, Scotland.*

Fold-mullions and Quartz-rods

A mullion is the vertical column which separates the lights of a window in the architecture of Gothic churches. If a whole series of stone mullions were placed side by side on the ground, they would then resemble *mullion structure*, which occasionally occurs, often in strongly deformed rocks. This structure often looks like a stack of telegraph poles or water-pipes, all lying parallel to one another, or it shows a resemblance to a rolled-up sheet of corrugated iron, as shown in **225**.

The striking appearance of mullion structure in the field arises not only as a consequence of its geometrical form, which has a markedly cylindrical character, but also as a result of its highly penetrative nature, affecting the whole structure of the rock in which it is developed. It is typically defined by bedding being folded, along with all the other structural elements in the rock, such as cleavage-planes, about a single axis, parallel to the plunge of the resulting *fold-mullions*.

The elongate columns forming the mullion structure are then bounded by bedding or cleavage-planes, which are commonly seen to be folded in a highly irregular manner when viewed in profile, at right angles to their plunge. However, it is often difficult to determine the exact nature of these surfaces, which are typically coated with a veneer of micaceous minerals, as shown in **226**. They often appear as merely a reflection of the internal fabric within the rock as a whole. This frequently occurs as a grain-fabric lineation, orientated parallel to the mullions themselves, particularly where elongate minerals are present in the rock.

Mullion structure is closely associated with folding. Indeed, fold-mullions are commonly seen at the contacts of layers differing from one another in competency, where they occur in the form of minor folds with markedly cuspate profiles. It also appears to be associated with pronounced stretching of the rock-mass in a direction parallel to the mullions themselves, thus emphasizing their cylindrical form. This stretching is likely to occur as a response to a constrictional deformation, rather than compression in a single direction.

Rodding is formed by elongate bodies of material, chiefly quartz, which has become segregated from its country-rocks during the course of the deformation, or has been introduced from elsewhere. Quartz-rods are formed as a result, as shown in **227**, varying considerably in size. Their cylindrical character may arise from the fact that they are often folded about an axis, lying parallel to their lengths even though this is not always the case by any means. Occasionally, it can be shown that the rodding is formed by quartz pebbles, which have been affected by extreme deformation of a constrictional nature, parallel to their lengths.

Figure 225 *Fold-mullions forming a set of markedly cylindrical structures, plunging at a low angle towards the left. The characteristic appearance of these mullions, resembling the stone columns used in the architecture of Gothic churches, results from the folding of bedding, along with any cleavages in the rock. Oykell Bridge (NC 385009), Sutherland, Scotland. (Field of view c 3.5m)*

Figure 226 *Fold-mullions in close-up, showing how the surfaces forming the individual mullions are often coated with a veneer of micaceous minerals. Coldbackie (NC 610601), Sutherland, Scotland.*

Figure 227 *Quartz-rods forming elongate segregations of vein quartz in the rock. It seems most likely that such quartz-rods are formed in response to a constrictional type of ductile deformation, drawing out these bodies parallel to the stretching-direction in the rock. Ben Hutig (NC 538653), Sutherland, Scotland.*

Development of Non-cylindroidal Folds

Minor folds typically show a wide variation in their geometrical features. Some folds are close to the ideal geometry of strictly cylindrical structures. For example, **228** shows a series of minor folds, all maintaining much the same plunge throughout a single exposure. The fold-hinges are effectively parallel to one another, so that each fold-closure maintains virtually the same profile as it is traced along its plunge. The structures formed as a result have a close resemblance to fold-mullions, taking only their external form into account.

By way of contrast, **229** shows a surface that has been folded in a much less regular manner, even though the folding still preserves a cylindroidal character, as marked by the rough parallelism which is shown by the fold-hinges. Such a degree of cylindroidal folding can be taken as rather more typical of minor folds in general. They are commonly found to be impersistent structures, with the individual fold-hinges replacing one another, as they each die out along their lengths. Such structures have been termed *pod-folds* from their resemblance to the shape of a pea-pod.

The variation in plunge that is typically a consequence of cylindroidal folding may be accentuated by the effects of further deformation. In particular, any stretching that occurs within the axial plane of a cylindroidal fold at a high angle to its fold-hinge tends to modify its original form. This effect is seen most clearly where such a fold has a hinge that is somewhat curved within its axial plane, as occurs in the case of pod-folds. The curvature shown by this fold-hinge then becomes accentuated during the course of any further deformation. If this effect is particularly pronounced, the cylindroidal character of the original folds may be entirely lost. This leads eventually to the formation of markedly non-cylindroidal structures, with fold-hinges that are very strongly curved within their own axial planes. They typically occur in the form of very pronounced domes and basins, with steeply-plunging terminations. A typical example is shown in **230**, where the highly irregular nature of the folded structure can be seen.

Such structures have been termed *sheath folds*, particularly where they are developed within ductile shear-zones as the result of very intense deformation, modifying the form of the original structures. It has been found that such structures may become so accentuated that the folding once more has an almost cylindrical appearance, as the fold-hinges rotate into virtual parallelism with the stretching-direction. However, ductile shear-zones are not the only environment in which these folds can occur, since they are also found wherever intense deformation results in stretching at a high angle to the regional fold-trend.

Figure 228 *Cylindrical folds affecting the bedding in a quartzite to form a series of fold-hinges, all plunging parallel to one another at a low angle towards the left. Glen Orchy (NN 243321), Argyll, Scotland.*

Figure 229 *Cylindroidal folds affecting a micaceous surface formed by the bedding in a metamorphic quartzite, showing how the fold-hinges are not all exactly parallel to one another, even though they still share the same general trend as one another. Loose block, Kinlochleven (NN 174616), Argyll, Scotland.*

Figure 230 *Non-cylindroidal folds, showing how the individual fold-hinges are often highly curved within the axial planes formed by these folds, so that they often plunge in diametrically opposite directions to one another. Port Cill Malluaig (NR 718697), Argyll, Scotland.*

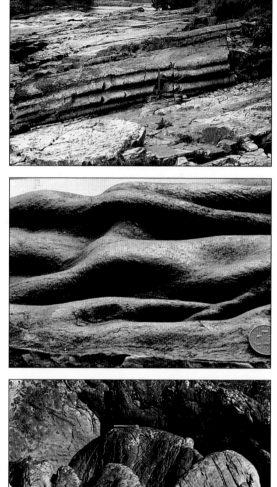

217

Evidence for Refolding

The highly deformed rocks forming the internal parts of orogenic belts are rarely affected by just a single phase of deformation and regional metamorphism. Instead, they often show evidence of refolding in response to a *deformation history* of more than a single phase. This evidence is best seen wherever two sets of minor folds are superimposed, one on top of the other, as shown in **231**. Note first the presence of a fairly tight fold with its axial plane dipping at a moderately steep angle towards the left. This fold clearly affects an earlier structure, best seen where it forms a very tight fold-closure towards the top of the photograph. The axial plane of this recumbent structure passes round the later fold-hinge, becoming much more steeply inclined as a result. The bedding affected by this earlier fold has clearly been refolded to form a more open structure, belonging to a later generation.

Often there are cleavages present in refolded rocks, axial-planar to folds of different generations. Indeed, the presence of a slaty cleavage implies that it was formed during the initial stages of any deformation history, since it is typically found axial-planar to folds which affect only the bedding. Such a cleavage and its associated folds are then known as first structures. For convenience, they are designated S_1 cleavages and F_1 folds, formed in response to the D_1 deformation. Bedding is then known as SS.

Likewise, the formation of a crenulation cleavage requires that an earlier fabric is present in the rock. Since this is most commonly formed by a slaty cleavage, it follows that crenulation cleavages typically occur in response to later phases in the deformation history. Such cleavages are usually found axial-planar to folds which affect any earlier cleavages as well as the bedding. These structures can then be designated S_2 cleavages axial-planar to F_2 folds, formed in response to the D_2 deformation; S_3 cleavages axial-planar to F_3 folds, formed in response to the D_3 deformation; and so on, according to the total number of deformation phases that can be recognized in any particular area.

232 shows a crenulation cleavage that is axial-planar to a series of rather upright folds. Although they only appear to affect the bedding, there is a small area towards the centre of the field of view, where the bedding is folded back on itself, forming the short and highly attenuated limb of an earlier fold. It is likely that the fabric affected by the crenulation cleavage is axial-planar to this fold.

Although crenulation cleavages are found axial-planar to folds of the same generation, they cut across the axial planes of any earlier folds, as shown in **233**, so allowing two distinct phases of deformation to be recognized in the rock. However, if such later cleavages are axial-planar to folds that affect the earlier structures, it is commonly found that an earlier cleavage or schistosity can best be recognized where it is folded around the hinge of these later folds, so providing another line of evidence showing that more than one phase of deformation has affected the rocks in question.

Figure **231** *Evidence of refolding, as shown by the presence of two sets of minor folds, superimposed one on top of the other as the result of two distinct phases in the deformation history affecting these rocks. Fraserburgh (NJ 999677), Buchan, Scotland.*

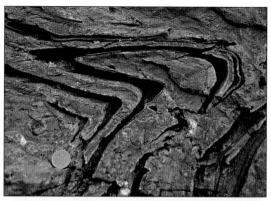

Figure **232** *Structural relationships showing evidence for two phases of deformation where a later set of upright folds with a crenulation cleavage parallel to their axial planes appears to affect an earlier structure, which is defined by the bedding being folded back on itself towards the centre of the photograph. Porthleven (SW 635248), Cornwall, England. (Field of view c 40cm)*

Figure **233** *Crenulation cleavage, dipping moderately towards the left in the darker bands, superimposed across a series of very tight fold-closures in the lighter layers, so providing evidence that two distinct phases of deformation have affected this rock. Trearrdur Bay (SH 249792), Anglesey, Wales.*

219

It is evidence of this sort that allows the dating of structures belonging to different generations, relative to one another. **234** shows several strain-bands as an example of the method, cutting across an earlier layering at a low angle, which are themselves affected by a later set of minor folds with their axial planes dipping at a moderate angle towards the left. The layering affected by the strain-bands may be designated S_1, the strain-bands S_2 and the later folds F_3.

However, it should be emphasized that crenulation cleavages tend to lose their distinctive character wherever they pass into coarser-grained schistosities, developed as a result of thorough recrystallization and the growth of new minerals under conditions of much higher metamorphic grade. Such schistosities look much like earlier schistosities in the field, formed in response to the D_1 deformation under similar conditions of regional metamorphism. The relationships just outlined are therefore particularly characteristic of low-grade metamorphic rocks.

Effect on Pre-existing Lineations The previous section has just dealt with the effects of refolding in two dimensions, since it is only the axial planes of the earlier folds which appear to register the results of the later movements. However, refolding typically occurs in three dimensions, generating a highly complex geometry for the folded layers. We therefore need to consider the orientation shown by the fold-hinges as well as the axial planes.

The first and most obvious affect of refolding on the orientation of any fold-hinge is that the early folds show systematic changes in plunge as they are traced around the later fold-hinges. This effect also extends to any earlier structure of a linear nature, such as bedding-cleavage intersections, stretching-directions, grain-fabric lineations, fold mullions, and so on. **235** shows a typical example where the intersection made by a crenulation cleavage with an earlier schistosity is itself affected by a later generation of structures, which also produces an intersection between this schistosity and a later set of crenulation cleavages. The latter intersection plunges towards the left at a low angle, while the former intersection plunges steeply down the dip of the now-folded schistosity. It can be seen that the earlier intersection, plunging steeply down-dip, is affected by the later folding of this schistosity about a more horizontal axis.

236 provides another example of refolding as seen in three dimensions. The hammer has been placed so that its shaft is parallel to the local plunge of a set of early fold-hinges, affecting the layer of light-coloured quartzite. These fold-hinges are themselves folded by a set of later folds with their axial planes dipping at a moderately steep angle towards the left.

Figure 234 *Strain bands affecting an earlier banding, which are themselves affected by a set of later folds, so providing evidence that three distinct phases of deformation have affected this rock. Trearrdur Bay (SH 249792), Anglesey, Wales.*

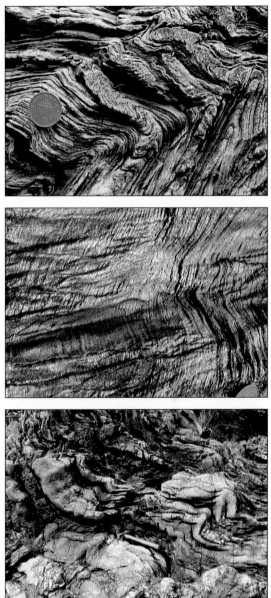

Figure 235 *Steeply-plunging intersection-lineation affecting a schistosity which is itself folded along with this lineation about an axis, plunging much less steeply towards the left, forming another intersection-lineation parallel to this direction. Loose block, Kinlochleven (NN 174616), Argyll, Scotland.*

Figure 236 *Early folds, plunging moderately towards the right, at least overall, which form the rather smooth surfaces to the quartzite layer, affected by a set of later folds with their axial planes dipping at a moderately steep angle towards the left. Port Ellen (NR 364448), Isle of Islay, Scotland.*

Development of Interference Patterns

While the plunge of an earlier fold-hinge is affected by any subsequent refolding, as just described, it is perhaps not so clear that the orientation of the later fold-hinges must be influenced by the geometry of the earlier folds. In fact, unless these earlier structures are isoclinal folds, the plunge of any later fold-hinges will vary according to the attitude of the layering to be folded, as it is traced around the earlier fold-hinges. In other words, not only does the plunge of an earlier fold change as it is traced across a later fold, but the plunge of such a later fold must also vary wherever it is superimposed across an earlier fold. It is the combination of these two factors which results in the formation of *interference patterns* between folds belonging to different generations.

Simple Domes-and-basins 237 shows the first type of interference pattern to be considered, which is defined simply enough by the presence of domes and basins, occurring in the form of structural culminations and depressions within the folded surfaces. These features can be recognized by the enclosed nature of the outcrops defined by these surfaces, as shown in close-up in 238.

These patterns are typically developed wherever the folding occurs on much the same scale, forming two sets of upright fold-structures, with steeply inclined axial planes, trending at a high angle to one another. A dome is formed wherever two antiforms, each belonging to a different generation, form a structural culmination at the point where they cross over one another. Likewise a basin is formed wherever two synforms, each belonging to a different generation, form a structural depression at the point where they cross under one another. The saddles separating these domes and basins from one another occur wherever an antiform is crossed by a synform, or vice versa. If such a folded surface were stripped bare by erosion, its shape would resemble an egg-box.

The geometrical form of this type of interference pattern makes it difficult to decide which are the earlier structures. What can be inferred is that two sets of fold-structures are present, each belonging to a different generation. They both pass through the horizontal as they cross one another, plunging away from the structural culminations, towards the structural depressions as a result.

Re-entrant Domes-and-basins 239 shows the second type of interference pattern to be considered. It is rather easier to interpret in terms of structural history, even though it shows a greater degree of structural complexity. 240 provides a closer view of the structural relationships. A simple dome or basin is no longer produced wherever antiforms or synforms, each belonging to different generations, cross over one another. Instead, the enclosed outcrops formed by the mutual interference of these two sets of fold-structures take on a crescentic form, with re-entrants present on only one side of the resulting culminations and depressions. It can be clearly seen that the later fold trends from side to side across the field of view, while the axial plane of the earlier fold swings through more than a right angle as it crosses the hinge of this later fold. It is this swing in strike, as shown by the earlier structure where it is affected by the later folding, which allows the relative ages of the two sets of fold-structures to be determined (cf. 238).

Figure 237 *Interference pattern formed by the superimposition of two sets of fold-structures to form a series of simple domes and basins in the surfaces affected by the folding. This pattern results wherever the two fold-sets have upright axial planes lying approximately at right angles to one another. Loch Monar (SH 197388), Inverness-shire, Scotland.*

Figure 238 *Close-up view of the interference pattern shown in the previous photograph, showing the enclosed nature of the outcrop pattern. It is difficult to date the two sets of fold-structures with respect to one another as the later folds do not affect the strike of the earlier axial-planes. Loch Monar (SH 197388), Inverness-shire, Scotland.*

Figure 239 *Interference pattern formed by the superposition of two sets of fold-structures to form a series of re-entrant domes and basins in the surfaces affected by the folding. This pattern occurs wherever a later set of upright folds affects an earlier set of folds with inclined axial-planes. Loch Monar (SH 197338), Inverness-shire, Scotland.*

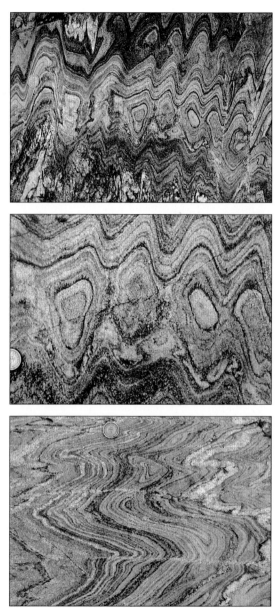

The earlier fold must change in plunge as it crosses over the crescentic-shaped outcrop shown in **240**. This means that it must plunge away from such an outcrop if it occurred in the form of a structural culmination, whereas it would plunge towards this outcrop if it formed a structural depression. The later fold by way of contrast closes in the same direction along its axial plane where it occurs on either side of this crescentic-shaped outcrop. This means that it must plunge in the same direction as far as both these areas are concerned. It can only do this if it first passes through the vertical and then the horizontal, or vice versa, as it is traced across the earlier folds. This can only happen if the earlier structures are a series of overfolds with their fold-limbs dipping in the same direction, while the later folds are still represented by rather upright structures. This type of interference pattern becomes even more complex if the earlier folds are so closely spaced that their outcrops start to coalesce with one another, as shown in **241**.

Zig-zag Refolds   **242** illustrates the third and final type of interference pattern to be considered. It shows the earlier fold-structures as trending across the field of view from upper left to lower right, while the later structures can be traced away from the camera at an oblique angle to this direction. Such an interference pattern is distinguished from the other two types by its lack of any enclosed shapes to the outcrop pattern, forming structural culminations and depressions within the folded surfaces. It has therefore exactly the same appearance as a refolded fold, where it is seen in profile, often as a vertical cross-section (cf. **231, 232** and **234**).

The earlier structures in this type of interference pattern show a constant sense of fold-closure as they are traced around the later fold-hinges. This means that the earlier folds do not pass through the horizontal to plunge in opposite directions as the result of the later folding.

However, the later fold-structures can also be seen to close in the same direction, wherever they are superimposed on top of the earlier folds. They share this feature in common with the second type of interference pattern, formed by re-entrant domes-and-basins. It means that the later fold-hinges pass first through the vertical and then the horizontal, or vice versa, as they are traced across the earlier folds, coming to plunge in the same direction, wherever they might be found. This means that the earlier folds are inclined structures with overturned limbs, while the later structures are rather more upright folds.

Figure 240 *Close-up view of the interference pattern shown in the previous photograph, showing how the culminations and depressions formed in the surface affected by the folding take on a crescentic form, marked by the presence of re-entrants in the outcrop pattern. Loch Monar (SH 197338), Inverness-shire, Scotland.*

Figure 241 *Complex interference-pattern of the same general type as shown in the last two photographs, formed wherever the earlier folds occur on a somewhat smaller scale than the later folds, giving rise to an outcrop pattern resembling the form of a mushroom in cross-section. Loch Monar (SH 197338), Inverness-shire, Scotland.*

Figure 242 *Interference pattern formed by the superposition of two sets of fold-structures, in which the original form of the earlier folds is retained as these folds are traced around the later fold-hinges, often developing a zig-zag pattern to the outcrops formed by the surfaces affected by the folding as a whole. Loch Monar (SH 197338), Inverness-shire, Scotland.*

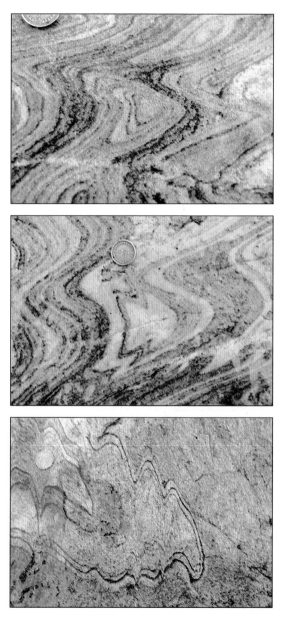

225

PART VI
BASEMENT ROCKS

Folded vein of granite pegmatite, lying across the foliation of a highly deformed gneiss. Torrisdale, Scotland.

Migmatites and Basement Gneisses

The structural features considered so far as the result of earth-movements have mainly affected sedimentary rocks or their metamorphic equivalents. These rocks were deposited for the most part as sedimentary sequences, lying unconformably on top of much older rocks. Often, this basement was caught up by the earth-movements responsible for folding and faulting now shown by the younger rocks, along with the effects of deformation and regional metamorphism. This is particularly the case where such a metamorphic basement and its sedimentary cover are seen within the confines of an orogenic belt. These basement rocks have usually been so profoundly modified by the orogenic movements that their original character is largely obscured. However, if they are traced away from the orogenic belt towards its foreland, it is found that they generally occur in the form of *basement gneisses*. Such rocks form the very foundations of the continents, emerging from below the unconformable sequences of younger sediments to form the Precambrian rocks of the continental shields. However, it may also be inferred that a similar basement of metamorphic rocks occurs throughout the continents at some depth below the surface, beneath an unconformable cover of sedimentary and other rocks.

The structural complexity typical of basement gneisses is partly a result of the variety of different rock-types from which they were originally formed. For example, *para-gneisses* are derived from rocks of a sedimentary nature, so that they often inherit a compositional banding which reflects the original presence of bedding in the rock. Metamorphic processes may also exert an important influence on the structure of such rocks, since they are formed under high temperature and confining pressure, as appropriate to very considerable depths within the earth's crust. Metamorphism under these conditions often does not appear to produce markedly planar fabrics, as found in regionally metamorphosed rocks of a much lower grade. Instead, such gneisses often occur in the form of rather massive rocks.

The great majority of basement gneisses appear to be derived from coarse-grained igneous rocks such as granites, along with various other rock-types of a more basic composition. Such rocks are generally known as *ortho-gneisses*. A characteristic feature exhibited by such rocks, particularly the more acid varieties, is the presence of a crude foliation, often marked by lines of recrystallized quartz and micas, winding between the feldspars. These foliated ortho-gneisses merge into *augen-gneisses* wherever these feldspar crystals, or aggregates of quartz and feldspar, occur in the form of lenticular "augen", wrapped around by the gneissic foliation in a stream-lined fashion (see **133**).

However, it is the common presence of granitic material in basement gneisses which perhaps exerts the most important influence on their structure, as seen in the field. This material typically occurs in the form of very thin and somewhat irregular sheets, interspersed in a discontinuous manner along the foliation of the original country-rocks, so forming what are often known as *lit-par-lit gneisses* (French: *lit-par-lit*, bed-by-bed). However, it may also penetrate the original country-rocks in a highly irregular fashion, often forming gradational contacts where it appears to merge with these rocks, while showing wide variations in grain-size, so that pegmatitic rocks are commonly found in such terrains. If there is a particularly intimate association between this granitic material and its host-rocks, the rock may be termed a *migmatite* (Greek: *migma*, a mixture).

It is often argued that the processes which come under the very general heading of *migmatization* are associated in some way with the presence of aqueous solutions or pore-fluids, possibly carrying silica and the alkalis in solution. The exact effect of such a fluid phase is a matter for some debate. For example, it may simply enhance the effects of *metamorphic differentiation*, whereby the various constituents in an otherwise solid rock can become segregated into separate layers, differing in their mineral composition from one another. Alternatively, some components of the original rock may go into solution, only to crystallize out elsewhere in the form of *segregation-veins* or *sweat-outs*.

Another possibility is that the pore-fluids are able to transport material in solution from deeper levels within the earth's crust, which then reacts with the country-rocks, transforming their chemical composition as the result of *metasomatic processes* (see p.40). However, if the temperature is sufficiently high, the presence of a fluid phase could equally well allow *partial melting* to occur, leading to the formation of a silicate melt of granitic composition, rich in volatiles. Finally, it is possible for such a granitic magma to be injected into its country-rocks from a source much deeper within the earth's crust, or even at greater depths within the upper mantle, giving rise to lit-par-lit gneisses without the intervention of a fluid phase at all.

Whatever their origin, **243** shows the typical appearance of lit-par-lit gneisses in the field. Note how the granitic layers, identified by their lighter colour, although lying along the foliation of the country-rocks, show a general lack of continuity. These layers are usually found to vary considerably in thickness and grain-size, while they often merge with more irregular masses of granitic material, isolated within the foliation of the surrounding rocks, as shown in **244**. A selvedge of darker rock, rich in micas, is commonly found along their margins. Another common feature is the presence of granitic veins, cutting obliquely across the layering, which otherwise forms the dominant structure in the rock. Lit-par-lit gneisses are often affected by folding, as shown in **245**, where mica-rich selvedges can also be seen, lying along the contacts of the granitic veins with the surrounding country-rocks.

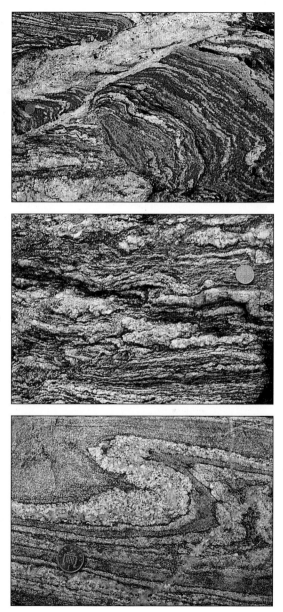

Figure 243 *Lit-par-lit gneisses showing features typical of migmatites in which irregular veins of granitic material penetrate their highly-metamorphosed country-rocks along the foliation. Cnoc Mhor (NC 757638), Sutherland, Scotland.*

Figure 244 *Migmatitic layering in a lit-par-lit gneiss, showing how the granitic material typically occurs in the form of irregular and somewhat discontinuous masses of light-coloured and rather coarse-grained rock, which lie along the foliation of the intervening country-rocks. Cnoc Mhor (NC 757638), Sutherland, Scotland.*

Figure 245 *Migmatitic gneiss in which the granite veins have become folded along with the surrounding country-rocks. Note the presence of dark-coloured selvedges to these veins, formed by a thin layer of mica-rich rock. Gearnasary (NC 740305), Sutherland, Scotland.*

Deformation Styles in Basement Gneisses

There are two distinctive styles of deformation which typically make their appearance in basement gneisses. Where deformation occurs under conditions of high temperature and confining pressure, perhaps enhanced by the presence of a fluid phase, it commonly results in a structural style that can only be described as highly plastic. It is often characterised by the disruption of different layers in the rock, particularly where these show considerable differences in lithology, accompanied by much folding of an apparently irregular nature. Such an appearance may however be misleading, at least to some extent, in that basement gneisses do not generally develop cleavages or other fabrics, which could otherwise be used to record the evidence of multiple phases of folding and deformation.

246 shows the typical appearance of lit-par-lit gneisses where folding has affected the layering to only a moderate degree. Such a fold-style is often described as *ptygmatic* (Greek: *ptygma*, folded matter), particularly where it affects granitic veins in what appears at first sight to be a highly irregular and tortuous fashion. There can be little doubt that such folds form as a result of buckling. They often adopt the form of *elasticas*, in which the folded veins are so contorted that they swing back and forth through more than 180° to form a series of closely-spaced loops (see **203**). By way of contrast, **247** shows a much more irregular style of deformation to affect migmatitic gneisses, illustrating a degree of structural complexity that might well be thought typical of basement gneisses in general.

Further complexity arises wherever there are marked differences in mineralogical composition between the various rock-bodies present in basement gneisses.

For example, where basic bodies are present in acid gneisses, these bodies are often penetrated by granitic veins, originating in the surrounding rocks. Eventually, these veins may expand and coalesce with one another, isolating fragments of the more basic rock as inclusions, surrounded by the more acid gneisses as shown in **248**. Such a rock is known as an *agmatite* (Greek: *agma*, a fragment). These inclusions are often gradually absorbed into the surrounding gneisses as a result of metasomatic processes taking place in the solid. A rock containing the partly-assimilated relics of these basic inclusions, often with very indistinct outlines, is known as a *nebulite* (Latin: *nebula*, a mist). Where flow has disrupted these inclusions in a highly irregular fashion as they were undergoing assimilation into the surrounding rock, they may only be seen in the form of wispy streaks of darker rock, known as *schlieren* (German: *schliere*, a streak or flaw in glass).

Such styles of highly-plastic deformation, as described in the previous paragraph, stand in complete contrast to the structural features of basement gneisses where they are cut across by ductile shear-zones. This commonly occurs as the result of *reworking* along the margins of a later orogenic belt, at such a depth that its roots are formed by a basement of older gneisses. This reworking often represents the initial stages in the development of a terrain of younger gneisses, flanked on one side by older gneisses, which were not themselves affected by the subsequent earth-movements. Traced into the younger terrain, this characteristic style of structural reworking is usually found to pass into a much more plastic type of deformation, similar to that already described, accompanying the formation of migmatitic gneisses of a new generation.

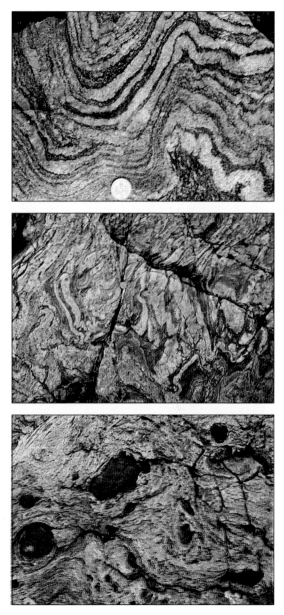

Figure 246 *Lit-par-lit gneisses showing the development of ptygmatic folds affecting a closely-spaced series of granitic veins. The nature of these folds suggests that the granitic veins acted as the more competent layers in the rock-mass as it underwent ductile deformation. Cnoc Mhor (NC 757638), Sutherland, Scotland.*

Figure 247 *Migmatitic gneisses in which the granitic portion forms a highly irregular network of folded veins and more diffuse patches of light-coloured rock, invading the darker country-rocks. Diabeg Pier (NG 800595), Wester Ross, Scotland. (Field of view c 1m)*

Figure 248 *Agmatite formed by the break-up of a more basic rock to form dark-coloured inclusions with highly irregular outlines, surrounded by much more acid gneisses. Note how the foliation in these gneisses appears to flow around these inclusions in a swirling fashion. Scourie Mor (NC 147458), Sutherland, Scotland. (Field of view c 1m)*

249 shows a typical example of a ductile shear-zone, developed on a small scale in banded gneisses. It crosses the field of view from top left to bottom right. The banding in the surrounding gneisses rotates into virtual parallelism with this shear-zone, and becomes attenuated to some extent, where it lies within its boundaries. Traced away from the area of older gneisses, such shear-zones gradually become wider and more closely-spaced, so that they eventually coalesce with one another parallel to the margins of the younger terrain. A characteristic feature of these areas in the presence of *banded gneisses* in which all the different elements are effectively parallel to one another, as shown in **250**.

Basic Dykes as Structural Markers
Determining the structural evolution of basement complexes is made much easier if basic dykes are present to act as structural markers, separating different episodes of deformation, regional metamorphism and migmatization from one another. Such a dyke is shown in **251**, dipping steeply towards the right. It has discordant contacts with the flat-lying foliation in the surrounding gneisses, so that its intrusion must have been the very latest event to affect these rocks.

Often, these dykes preserve chilled contacts with their country-rocks, while they commonly retain their primary textures, characteristic of igneous rocks in general, even if their original mineralogy has been altered to some extent. These dykes usually occur in the form of dyke-swarms, maintaining much the same orientation over quite wide areas. Evidently, such areas of basement gneisses have not been affected by any tectonic and metamorphic activity, following the intrusion of these dykes.

Traced into a terrain affected by later movements, these dykes become deformed and metamorphosed along with their surrounding country-rocks. As far as the latter are concerned, this is commonly associated with a change in the metamorphic grade displayed by these gneisses, coupled with the development of new structures in the rocks. Often, these rocks are so reconstituted by the renewed earth-movements that they must be regarded as virtually new rocks, particularly where the later movements are accompanied by migmatization and the intrusion of much granitic material.

The effect of these later events on any dykes that can be used as structural markers in basement terrains is first seen in the gradual loss of their igneous textures as these rocks become increasingly schistose. At the same time, these dykes become so deformed that they no longer clearly display any intrusive contacts with their country-rocks, while they gradually lose any evidence of their original form as igneous intrusions. Eventually, they become incorporated into the reworked gneisses of the younger terrain in such a way that they have effectively lost any sense of their separate identity as igneous intrusions.

Figure **249** *Ductile shear-zone cutting a series of banded gneisses. Note how the banding of these gneisses is deflected in a sigmoidal fashion as it crosses this shear-zone, eventually coming to trend almost parallel to its boundaries along the length of the shear-zone iteslf. Rhicarn (NC 076255), Sutherland, Scotland. (Field of view c 2m)*

Figure **250** *Banded gneisses, formed most likely as a result of extreme deformation, which has caused the various layers now seen in the rock to rotate into parallelism with one another, so modifying whatever structure these rocks originally displayed. Annagh Head (F 640340), County Mayo, Ireland.*

Figure **251** *Basic dyke forming a parallel-sided body of much darker rock, which cuts discordantly across the flat-lying foliation in the acid gneisses that form its wall-rocks on either side of this intrusion. Achmelvich Bay (NC 059243), Sutherland, Scotland. (Field of view c 2m)*

252 shows a basic dyke where it has been folded in response to later earth-movements, during the initial stages in the reworking of a basement complex. Note that this intrusion still preserves a certain degree of structural discordance with the foliation in the surrounding gneisses, best seen where its contacts with these rocks are close to the vertical.

However, these basic dykes and their country-rocks much more frequently come under the influence of ductile shear-zones as they are traced into the younger terrain. The intense deformation occurring within such a shear-zone typically reduces the structural discordance between these dykes and the surrounding rocks to virtually zero, often in a very abrupt manner. This can only happen if the displacements taking place across these shear-zones are so large that all the structural elements in the rock rotate into parallelism with one another, whatever might have been their original attitude. This means in effect that any basic dykes originally present as discordant intrusions eventually come to form concordant layers in a banded gneiss, sharing the same foliation as the surrounding rocks. Once these shear-zones start to merge with one another, a new structural trend is defined, typical of the younger terrain.

Occasionally, as shown in **253**, the discordant nature of the original dykes can still be discerned, despite all the effects of structural reworking. The thin layer of dark rock which crosses the field of view from top left to lower right is the remnants of an early dyke, still showing a degree of discordance to the foliation of the surrounding gneisses. It can be seen to cut across a discontinuous band of basic gneiss immediately below the lens-cap, which reappears on its other side, much further to the left. **254** shows by way of contrast the more typical appearance of these basic bodies where they form concordant layers of darker rock, completely incorporated into the structure of the surrounding gneisses. Often, these sheets start to break up, so forming the basic inclusions which are commonly seen in more acid gneisses (see **248**).

Figure 252 *Folded dyke, forming an irregular body of darker rock, surrounded by more acid gneisses, affected by the early stages in the reworking of a basement complex. Achmelvich Bay (NC 053260), Sutherland, Scotland.*

Figure 253 *Thin sheet of basic rock, cutting discordantly across the structure of the surrounding gneisses at a very oblique angle, so retaining some evidence that it was originally intruded into these rocks as a dyke. The gneisses have therefore been affected by a considerable degree of reworking, following its intrusion. Laxford Brae (NC 235490), Sutherland, Scotland.*

Figure 254 *Banded gneisses in which the dark-coloured layers of basic rock show no sign of any structural discordance even though these layers probably represent dykes that were once intruded across the structure of the surrounding gneisses, prior to reworking affecting the whole complex. Ceannabeine (NC 442655), Sutherland, Scotland.*

Appendix I: Guidelines for Structural Fieldwork

All structural fieldwork ought to be undertaken within the context of an accurately-constructed geological map at an appropriate scale for the area under study. Indeed, the construction of such a map is often the prime objective of the exercise, perhaps leading to more detailed studies of specific aspects of the geology at a later date. However, the student would do well to appreciate from the outset that there is a certain dichotomy in approach to geological fieldwork, since any geological map can only be produced by outcrop mapping, whereby the geological boundaries are traced out by following them in the field, and then transferring these lines to a base map. More detailed studies of the geology require by way of contrast a much closer examination of individual exposures, to determine the characteristic features of the rocks themselves, even though it is obviously this evidence which is used to locate the boundaries on the geological map in the first place.

It is always a good idea to view any exposure at first from a distance, since this often reveals more clearly the gross structure of the rocks than a closer view. The exposure can then be approached, and its features examined in more detail. The lithology and texture of the rocks forming the exposure should first be noted down in the field notebook. If more than a single lithological unit is present, the nature of the geological contacts between these different units should be determined, bearing in mind that these contacts may well represent a boundary on the geological map.

In dealing with sedimentary rocks, it may be possible to log a vertical section through these rocks as part of the stratigraphic sequence for the area as a whole, particularly if exposure is more or less continuous. A search should also be made for any fossils in suitable horizons. If the exposure is formed by igneous rocks, a search should first be made for any contacts that are present between different intrusions, and any evidence of their age-relations determined. Once these general observations have been made, the structural features of the rocks can be considered under the following headings, taking structural measurements wherever possible, and plotting them as appropriate on the geological map. It is assumed that the reader is familiar with the application in structural geology of the stereographic projection, based on the use of an *equal-area net*.

Bedding

The dip and strike of the bedding should be measured wherever possible at each exposure, and the observation plotted on the geological map, using a strike bar with a tick on one side to show the direction of the dip, together with a number giving the angle of dip in degrees. What is loosely known as a "dip arrow" should not be used to record observations of what is effectively a planar element of the geological structure. Instead, such arrows should only be used on a geological map to show the plunge of linear features, such as fold-hinges and intersection-lineations. Wherever possible, sedimentary structures should be used to determine whether or not the bedding is right-way-up. A special symbol is often used on the geological map to show where the bedding is inverted. The stereographic projection can then be used to plot these observations as poles to the bedding-planes on what is known as a pi-diagram. This will show if the bedding is folded about a common axis, corresponding to the plunge of any cylindroidal folds which are developed within the particular area under consideration.

Jointing

Many exposures show complex patterns of jointing, which can only be recorded adequately by taking a representative sample, numbering perhaps a hundred joint-planes, measuring the dip and strike of each plane, and plotting the observations on the stereographic projection in the form of a pi-diagram, showing the poles to the joint-planes. However, this procedure is much simplified wherever flat-lying sediments are cut by sets of vertical joints. Although a representative sample still needs to be made, it is then only necessary to measure the strike of each joint to determine its orientation. The observations can then be plotted on a rose diagram, showing their frequency of occurrence as a function of their strike. Particular attention should be paid to the relationship of these joints to any other structures in the rock, since this may reveal whether particular joint-sets occur in the form of extension-fractures or shear-fractures.

Faulting

Many faults are only revealed as the result of geological mapping, since they are not exposed anywhere at the surface. It is often very difficult to determine the dip of such a fault, unless its outcrop is clearly affected by the topography. This means that it is not always possible to determine the character of a particular fault, except as a matter of inference. Often, only the up-thrown or downthrown sides of such faults are marked on the geological map, presumably on the assumption that strike-slip movements are not involved. However, if a fault-plane is exposed in the field, its dip and strike can be measured directly. The orientation of any striations developed on the fault-plane should also be determined. This is best done by measuring the angle between the direction defined by these striations and the strike of the fault-plane itself. This gives what is known as the *pitch* of the striations within the fault-plane, rather than their absolute plunge.

Minor Folding

Much structural fieldwork is directed towards the mapping of major folds through the detailed observations of what are known as minor structures. Wherever folding is seen on a minor scale in the field, the plunge of the fold-hinges should first be measured at each exposure, and the observations plotted on the geological map, using plunge arrows. If these folds lack an axial-planar cleavage, the orien-

tation of their axial planes should be measured directly in the field. Next, the surface affected by the minor folding should be examined, to determine whether it affects just the bedding, or whether an earlier cleavage is folded alongwith the bedding, as the result of more than a single phase of deformation. If a cleavage is present, axial planar to these folds, its dip and strike should also be measured, preferably along the axial plane of a particular fold. These measurements also need to be plotted on the geological map, using the appropriate dip-and-strike symbol. The nature of this cleavage should also be recorded, since a slaty cleavage is often developed axial planar to the earliest F_1 folds, while crenulation cleavages are typically found along the axial planes of the subsequent F_2 and later folds, at least in rocks affected by a regional metamorphism of relatively low grade. The dip-and-strike symbols used to plot these observations on the geological map should clearly distinguish between these different types of axial-planar cleavages. Finally, the style of the minor folds should be recorded with particular reference to the sense of asymmetry shown by these structures. The arrows showing the plunge of the minor fold-hinges on the geological map can be annotated to show whether these minor folds appear S-shaped or Z-shaped when they are viewed down-plunge, so allowing the axial planes of the major folds to be located in the field. All these structural observations should be treated stereographically, and the resulting diagrams compared with the corresponding pi-diagram for the same area, showing the poles to the bedding.

Cleavage and Lineation

Even if the bedding is not folded within the confines of a single exposure, it may be cut by a cleavage which is axial planar to folds on a larger scale. The orientation of this cleavage should be measured in terms of its dip and strike at each exposure, and the observations plotted on the geological map, using the appropriate symbols. The plunge of its intersection with the bedding should also be measured, and then plotted on the geological map, using plunge arrows. The relationship of the cleavage to the bedding also needs to be recorded at each exposure, and this information used to locate the axial planes of the major folds in the field. Finally, the plunge of any lineations associated with these or other cleavages should be measured, and the observations plotted on the geological map, paying particular attention to the possible development of a stretching-direction in the rocks under study. All these structural observations should also be plotted on the appropriate stereogram, showing the plunge of the minor fold-hinges and the orientation of the cleavages axial-planar to these minor folds, as described above.

Structural Facing and Stratigraphic Sequence

Where the stratigraphic order of the rocks affected by all these minor structures can be determined through the use of sedimentary structures, it is also possible to determine a corresponding direction of structural facing. Although this direction is an important element in any structural interpretation on a regional scale, the facing of minor structures can also be used as evidence for refolding on a much more local scale. For example, the facing shown by earlier structures will change wherever they are affected by later structures, while the facing shown by later structures will likewise change wherever they are superimposed across earlier structures. Sedimentary structures can also be used to establish the stratigraphic sequence for the area under consideration, particularly where fossils are either lacking or poorly preserved. However, care needs to be exercised, especially in areas of structural complexity. Much the most useful information concerning the nature of the major structures, as well as the stratigraphic relationships of the rocks as a whole, then comes from the immediate vicinity of the formation boundaries, wherever these can be shown to be stratigraphic in their character. Ideally, the stratigraphic relationships can be established by determining the direction of younging across such a boundary through the use of sedimentary structures as indices of stratigraphic order, assuming that it can also be demonstrated that the structural relationships remain constant across this boundary as well. Under these circumstances, it becomes possible to reconstruct a stratigraphic sequence for the whole area, despite the complexity of the geological structure.

Appendix II: List of Field Localities

The localities shown in this book have been listed in alphabetical order. National Grid Reference Numbers have been given to help locate the geographical position of each locality precisely.

Aberdeenshire

Collieston (NK 060308): **219** recumbent fold (facing)

Fraserburgh (NJ 999677): **231** refolded fold

Quarry Head (NJ 898651): **109** buried landscape

Spar Craigs, Belhelvie (NJ 933197): **24** mimetic grading

Anglesey

Newborough Warren (SH 391634): **82** and **84** pillow lava; **83** lava pillow

Rhoscolyn (SH 264749): **172** flattened folds; **198** and **199** spaced cleavage

South Stack (SH 205823): **216** axial-planar cleavage

Trearddur Bay (SH 249792): **233** and **234** superimposed strain-bands; (SH 251791): **177** attenuated fold-limbs; **207** and **208** strain-bands

Angus

Stannergate, Dundee (NO 443310): **81** Neptunian dyke

Argyll

Aonach Dubh (NN 139569): **70** lava-flows

Ballachulish (NN 024595): **95** xenoliths; **99** sheeting

Buchaille Etive Mor (NN 227547): **86** intrusive contact; **87** vent margin

Caolasnacon (NN 142607): **12** cross-bedding

Coire nam Beith (NN 139553): **77** auto-brecciation

Crarae (NR 996981): **94** igneous dyke

Easdale (NM 748171): **204** intersection-lineation

Glen Orchy (NN 243321): **19** disharmonic flow-folds; **228** cylindrical fold-hinges

Kentallen (NN 011582): **90** cross-cutting intrusions

Kilmory Bay (NR 697748): **217** bedding–cleavage intersection

Kinlochleven (NN 174616): **229** cylindroidal folds; **235** superimposed intersection-lineations

Port Cill Malluaig (NR 718697): **230** non-cylindroidal folds

Banffshire

Banff (NJ 682647): **37** channel margin

Knock Head (NJ 661659): **22** graded bedding

Berwickshire

Grantshouse (NT 811625): **114** plumrose markings; **135** fault-vein fibres

Siccar Point (NT 813710): **100** angular unconformity; **101** erosion surface; **107** overlap

Caithness

Clairdon Head (ND 138700): **60** desiccation cracks

South Head (ND 376497): **4** bedding-plane; **62** and **63** synaeresis cracks

Charnwood Forest, Leicestershire

Beacon Hill (SK 510148): **115** en échelon fringes; **192** bedding–cleavage relationships

Bradgate Park (SK 529112): **201** cleavage refraction

Cornwall

Bude (SS 201073): **18** climbing ripples; **51** flame structure

Cracklington Haven (SS 139966): **46** and **47** groove-casts; **212** S-shaped fold-pair; (SS 139968): **139** feather-fractures; (SS 139968): **218** bedding–cleavage relationships

Duckpool (SS 201116): **117** extension fibres

Longcarrow Cove (SW 893768): **203** divergent cleavage-fan

Millock Haven (SS 186006): **165** chevron folds; **220** structural facing

Porthleven (SW 635248): **163** parallel fold; **169** and **170** buckle-folds; **171** composite similar folds; **205** crenulation cleavage; **232** superimposed folds

Rinsey Head (SW 593269): **85** igneous contact

Trebarwith Strand (SX 048865): **117** mineral vein; (SX 051867): **158** décollement

Widemouth Bay (SS 198035): **5** folds; **46** groove-casts; **50** load-casts; (SS 196014): **44** bounce-marks; **45** prod-marks; (SS 194013): **120** sigmoidal tension-gashes; (SS 197036): **181** incipient boudinage; (SS 197038): **141** normal faults

Cumbria

Kentmere (NY 459074): **195** and **196** bird's eye tuff

Salton Bay (NX 958158): **103** erosion surface

Shap (NY 556084): **96** xenolith

Tebay (NY 607010): **160** feather fractures; **161** en échelon tension-gashes

Devon

Baggy Point (SS 426401): **56** slump-sheet; (SS 435401): **119** en échelon tension-gashes

Hartland Quay (SS 224247): **43** longitudinal scour-marks; **49** load-casts

Torquay (SX 922628): **35** colonial coral

Durham County

High Force (NY 880284): **125** shatter-zone

Dyfed

Aber-arth (SN 490649): **2** bedding

Abereiddy (SM 801321): **88** chilled margin

Amroth (SN 156068): **54** ball-and-pillow

Broad Haven (SM 860142): **54** ball-and-pillow; **138** listric fault; **144** break-thrust; **162** extension veins

Clarach (SN 587836): **27** distal turbidities

Cwmtudu (SN 359580): **200** spaced cleavage; (SN 359580): **202** convergent cleavage-fan

Glossary

Geological terms that appear in the text but that are not listed in the Index are explained here.

acid rocks Any *igneous rock* containing more than 66% silica, SiO_2 or thereabouts, typical of *granites* and their finer-grained equivalents.

agglomerate A *pyroclastic rock* in which fragments of volcanic and other rocks lie in a matrix of finer-grained *ash* or *tuff*, produced by the force of volcanic explosions.

amphibole A complex group of rock-forming minerals, consisting of hydrous silicates of magnesium, iron, calcium, sodium and aluminium. Commonly found in igneous and metamorphic rocks.

amygdaloidal texture Characteristic of fine-grained *igneous rocks* in which vesicles have been filled by secondary minerals at a late stage in the consolidation of the magma.

andalusite A mineral occurring as one form of the aluminium silicate Al_2SiO_5, commonly found in rock thermally metamorphosed at low confining pressures.

anhydrite A mineral formed by calcium sulphate, $CaSO_4$.

ankerite A mineral formed by the double carbonate of calcium and iron, $CaFe(CO_3)_2$.

arkose A sandstone containing much detrital *feldspar* as well as *quartz*, chiefly in the form of *microcline*, which gives a pinkish colour to the rock.

ash The name applied to any fine-grained *pyroclastic* material erupted as fragments by the force of volcanic explosions.

barytes A mineral formed by barium sulphate, $BaSO_4$.

basalt A fine-grained *igneous rock* of basic composition, composed of calcic *feldspar*, *pyroxene* and iron ore as essential minerals, together with interstitial glass in some examples.

basic rocks Any *igneous rock* containing between 45% and 52% SiO_2 or thereabouts, typical of *basalts* and their coarser-grained equivalents.

biotite A dark *mica* forming a hydrous alumino-silicate of potassium, magnesium and iron, commonly found in *igneous* and *metamorphic rocks*.

breccia A *sedimentary rock* consisting of angular fragments of pre-existing rocks over 2mm in diameter, generally set in a finer-grained matrix of *detrital* material.

calcite A mineral formed by calcium carbonate, $CaCO_3$.

chalcedony A *cryptocrystalline* form of silica, consisting mostly of *quartz*, together with a certain amount of water.

chert A siliceous rock composed predominantly of crystalline *quartz*, extremely fine-grained, which occurs in the form of bedded deposits.

chlorite A group of minerals structurally similar to the *micas*, forming hydrous alumino-silicates of aluminium, iron and magnesium, commonly found in low-grade *metamorphic rocks*.

clastic limestone The name applied to limestones formed by fragmental material such as organic debris (shells, etc.) or any other pre-existing carbonate rock.

clay minerals A group of minerals, generally occurring as very fine-grained particles, consisting of hydrous alumino-silicates, mainly of aluminium and magnesium, showing a layered structure like the *micas*, and capable of absorbing large amounts of water.

cleavage The way in which a rock or mineral parts or splits.

conglomerate A *sedimentary rock* consisting of rounded fragments of pre-existing rocks over 2mm in diameter, consisting of pebbles, cobbles or boulders, set in a finer-grained matrix of *detrital* material.

contact metamorphism Alteration of pre-existing rocks by heat alone. Typically affects the country rocks around an igneous intrusion.

cordierite A mineral commonly found in *metamorphic rocks*, forming an alumino-silicate of aluminium, iron and magnesium.

country-rocks The rocks immediately in contact with an intrusive mass of igneous rock.

cryptocrystalline A very fine-grained crystalline texture in which individual crystals can only be detected using a powerful microscope.

detrital A term applied to any particles which have been derived from a pre-existing rock.

dolerite An *igneous rock* forming the hypabyssal equivalent of basalt, consisting of calcic *feldspar*, *pyroxene* and iron ores as its essential minerals. Known as diabase in USA and Europe.

dolomite A mineral formed by the double carbonate of calcium and magnesium, $(CaMg)CO_3$.

dolomite A sedimentary rock consisting predominantly of the mineral dolomite rather than calcite.

eclogite A *metamorphic rock* consisting essentially of *pyroxene* and *garnet*, formed from *basic rocks* under conditions of extremely high pressure.

evaporites *Sedimentary rocks* formed in general by the precipitation of minerals from aqueous solutions such as sea-water, forming deposits of rock-salt, *gypsum* and *anhydrite*.

explosion breccia A volcanic *breccia* formed virtually *in situ* by the explosive shattering of the country-rocks in the vicinity of a volcanic vent.

feldspar The most important group of rock-forming minerals, consisting of alumino-silicates of potassium (orthoclase and *microcline*), sodium (albite) and calcium (anorthite), with the latter two minerals forming the end-members of a *solid solution series* known as the plagioclase feldspars.

flint A siliceous rock composed predominantly of microcrystalline *quartz*, which typically occurs in the form of nodules, concretions and other types of segregation in sedimentary rocks.

foliation Any planar structure formed by the parallel orientation of platy minerals in a rock, particularly where this is emphasised by the segregation of different minerals to form lithological layering on a small scale.

garnet A complex group of minerals, mostly found in *metamorphic rocks*, forming silicates of aluminium, iron, manganese, chromium, calcium and magnesium.

gneiss A *metamorphic rock* of coarse grain-size greater than 2mm, which generally lacks a *schistosity*, although a foliation may be present, forming a lithological layering in the rock.

granite The name used collectively for any coarse-grained igneous-looking rock of acid composition, consisting of *quartz*, potash feldspar, and some sodic feldspar, together with minor amounts of dark minerals such as *biotite* or *hornblende*.

greywacke A *sandstone* consisting predominantly of rock and mineral fragments, set in a finer-grained matrix rich in *clay minerals*.

gypsum A mineral formed by hydrated calcium sulphate, $CaSO_4$. $2H_2O$.

haematite An ore mineral formed by iron oxide in the ferric state, Fe_2O_3.

hornblende A mineral of the *amphibole group*, forming a hydrous alumino-silicate of aluminium, iron and magnesium, together with minor amounts of sodium and calcium, commonly found in *igneous* and *metamorphic rocks*.

hornfels A hard and splintery rock produced by the thermal alteration of *shale* or *mudstone* as a result of *contact metamorphism*.

hypabyssal rocks Any *igneous rock* of medium grain-size, which crystallized at relatively shallow depths within the earth's crust, often in the form of minor intrusions.

igneous rocks Rocks which have solidified from molten magma. One of the three main groups of rocks which make up the material of the earth's surface (see *sedimentary* and *metamorphic rocks*).

ignimbrite A *pyroclastic rock* formed by the eruption of an ash-flow, typically consisting of crystals and rock-fragments lying in a matrix of glass-shards, which may be welded together into a compact mass.

intermediate rocks Any *igneous rock* containing between 52% and 66% SiO_2 or thereabouts.

intrusion breccia A volcanic breccia in which the fragments are not all of local derivation, in contrast to an *explosion breccia*.

kyanite A mineral occurring as one form of the aluminium silicate, Al_2SiO_5, commonly found in *schists* as a product of *regional metamorphism* under lower temperatures but higher pressures than *sillimanite*.

limestone A *sedimentary rock* composed predominantly of *calcite*.

limonite A general name for all the oxidation and hydration products of iron-bearing minerals, consisting of various amorphous and *cryptocrystalline* constituents, mainly iron oxides and hydroxides, predominantly yellow or brown in colour.

marble A metamorphic *limestone* in which the calcite and other minerals have suffered re-crystallization.

metamorphic rocks Rocks which were originally igneous or sedimentary but which have been changed in character or appearance by heat, pressure or fluids. One of the three main groups of rocks which make up the material of the earth's surface (see *igneous* and *sedimentary rocks*).

mica A complex group of minerals, consisting of hydrous alumino-silicates of aluminium and potassium, together with iron and magnesium in the darker varieties such as *biotite*, showing a crystal structure with a perfect basal cleavage.

microcline A potassium feldspar ($KAlSi_3O_8$) typically found in potassic granites, as well as sedimentary *arkoses* derived by the weathering and erosion of such rocks.

microcrystalline A fine crystalline texture in which individual crystals cannot be resolved by the naked eye but which can be detected using a microscope.

mudstone A *sedimentary rock* consisting predominantly of *clay minerals* less than 1/256mm in diameter, which lacks the fissility of a *shale*.

muscovite A white *mica* forming a hydrous alumino-silicate of potassium and aluminium, commonly found in *schists* and *acid rocks*, and as a *detrital* mineral in *sandstones*.

olivine A group of rock-forming minerals varying in composition between the silicates of iron and magnesium, commonly found in *basic* and *ultrabasic rocks*.

orthoclase The common form of potash feldspar ($KAlSi_3O_8$), typically found in *granites* and other *acid rocks*.

pegmatite Any igneous-looking rock of particularly coarse grain, often granitic in its composition, and typically found in association with *plutonic rocks* of finer grain.

phenocrysts See *porphyritic texture*.

phyllite A *metamorphic rock* lacking the well-developed foliation of a *schist*, but yet in a more advanced state of recrystallization than a *slate*.

phyllonite A *metamorphic rock* looking much like a *phyllite*, formed as a result of *retrograde metamorphism* affecting the mineral constituents of its coarse-grained progenitor.

plagioclase feldspars A series of soda-lime feldspars varying continuously in composition from albite ($NaAlSi_3O_8$) to anorthite ($CaAl_2Si_2O_8$). Commonly found in *igneous rocks*, particularly those of intermediate or basic composition.

plutonic rocks Any *igneous rock*, generally with a grain-size greater than 5mm, resulting from its crystallization at some depth within the earth's crust, often in the form of major intrusions.

porphyritic texture Characteristic of *igneous rocks* containing relatively large crystals, known as *phenocrysts*, set in a matrix of finer-grained or glassy rock.

porphyroblasts Larger grains, often with well-developed faces, set in the finer-grained matrix of a *metamorphic rock*.

pyrites A mineral formed by iron sulphide, FeS_2.

pyroclastic rocks Rocks made up of fragmental volcanic material produced in volcanic eruptions.

pyroxene A complex group of rock-forming minerals, consisting of various silicates of iron and magnesium, together with calcium, sodium and aluminium in some varieties, commonly found in *basic* and *ultrabasic rocks*.

quartz A common rock-forming mineral formed by silica, SiO_2.

quartzite Strictly, the name applied to a *metamorphic rock* formed by the recrystallization of a *quartz*-rich *sandstone*, but also used to describe quartz-rich sandstones in which the *detrital* grains are themselves set in quartz cement.

regional metamorphism The alteration of pre-existing rocks in response to changes in temperature and pressure; typically affects rocks undergoing deformation within orogenic belts.

retrograde metamorphism Any change in lithology which results from the lowering of temperature or confining pressure during the course of *regional metamorphism*.

salt deposits See *evaporites*.

sandstone A *sedimentary rock* consisting predominantly of *detrital* grains between 1/16mm and 2mm in diameter, usually formed by *quartz* but also including any other mineral grains or rock fragments.

schist A *metamorphic rock* with a schistose texture, formed by the parallel orientation of platy minerals such as the *micas*.

schistosity The platy structure characteristic of schists, and formed as a response to regional metamorphism.

sedimentary rocks Any rock deposited at the earth's surface as the result of external processes, usually forming a stratified sequence of layers or beds. Made up of material derived from pre-existing rocks, organic matter and/or chemical/biochemical precipitates. One of the three main groups of rocks which make up the material of the earth's surface (see *igneous* and *metamorphic rocks*).

sericite The name given to any fine-grained and colorless micaceous mineral, similar in composition to *muscovite*.

shale A *sedimentary rock* consisting predominantly of *clay minerals* less than 1/256mm in diameter, formed by the consolidation of a mud, which differs from a *mudstone* in its fissile character, splitting easily along the bedding.

siderite A mineral formed by iron carbonate, $FeCO_3$.

sillimanite A mineral occurring as one form of the aluminium silicate, Al_2SiO_5, commonly found as a product of *regional metamorphism* under higher temperatures but lower pressures than *kyanite*.

siltstone A *sedimentary rock* composed predominantly of *detrital* material varying in grain-size between 1/256mm and 1/16mm.

slate Any fine-grained rock which splits easily into thin slabs oblique to the bedding, but more particularly such rocks formed originally from *shales*, *mudstones* and volcanic ashes as a result of their deformation.

solid solution series The substitution of one element for another in a crystal lattice, commonly found in silicate minerals such as *feldspars, amphiboles, pyroxenes* and *olivines*.

spotted slate A slaty rock in which there are a conspicuous series of small spots, marking the incipient formation of *porphyroblasts* as the result of *contact metamorphism*.

tuff The consolidated equivalent of a volcanic ash, forming a fine-grained *pyroclastic rock*.

ultrabasic rocks Any *igneous rock* with less than 45% SiO_2, characteristic of rocks containing little or no *feldspar*.

vesicular texture Characteristic of fine-grained *igneous rocks* in which cavities are present, formed by gas-bubbles.

volcanic rocks Any *igneous rock*, generally with a grain-size of less than 1mm, which has been erupted from a volcano.

welded tuff A *pyroclastic rock* formed by the welding together of glass-shards and other fragments while they still remain hot and therefore plastic after their eruption from a volcano.

List of Selected References

Basic Geological Mapping, by John Barnes. Geological Society of London Handbook, Open University Press, 1981.
Challinor's Dictionary of Geology (6th edition), edited by Antony Wyatt. University of Wales Press, 1987.
Dictionary of Petrology, by S.I. Tomkieff (edited by E.K. Walton, B.A.O. Randall, M.H. Battey and O. Tomkieff). John Wiley and Sons, 1983.
Field Description of Igneous Rocks, by Richard Thorpe and Geoff Brown. Geological Society of London Handbook, Open University Press, 1985.
Field Description of Metamorphic Rocks, by Norman Fry. Geological Society of London Handbook, Open University Press, 1984.
Field Description of Sedimentary Rocks, by Maurice E. Tucker. Geological Society of London Handbook, Open University Press, 1982.
Field Geology (6th edition), by Frederick H. Lahee. McGraw-Hill, 1961.

Geology in the Field, by Robert R. Compton. John Wiley and Sons, 1985.
Introduction to Geological Maps and Structures, by John L. Roberts. Pergamon Press, 1982.
Introduction to Small-Scale Geological Structures, by Gilbert Wilson. George Allen and Unwin, 1982.
Mapping of Geological Structures, by K. McClay. Geological Society of London Handbook, Open University Press, 1987.
Sedimentary Structures, by J.D. Collinson and D.B. Thompson. George Allen and Unwin, 1982.
Sequence in Layered Rocks, by R.R. Shrock. McGraw-Hill, 1948.
Techniques of Modern Structural Geology (Volume I: Strain Analysis), by John G. Ramsay and Martin I. Huber. Academic Press, 1984.
Techniques of Modern Structural Geology (Volume II: Folds and Fractures), by John G. Ramsay and Martin I. Huber. Academic Press, 1987.

Subject Index